Rolf H. J. Schlegel, PhD

Concise Encyclopedia of Crop Improvement
Institutions, Persons, Theories, Methods, and Histories

Pre-publication
REVIEWS,
COMMENTARIES,
EVALUATIONS . . .

"This book offers a concise history of crop development. It is a handy reference for knowing our past and understanding our present. It is written by one of the few modern scientists who has lived with and met many of the people in his book or those who knew them. He has done us a great service by writing this concise history before it is lost in the momentum of the ever-expanding current research in plant breeding. One notable aspect of this book is that it is written from a European perspective, hence is particularly valuable for American readers who are well versed in our traditions, but have not always seen how global science interacts."

P. Stephen Baenziger, PhD
Eugene W. Price Distinguished Professor,
University of Nebraska – Lincoln

"Professor Schlegel has done a great service by assembling voluminous data on historical aspects of crop development. His treatment calls attention to the evolution of methods for plant breeding as well as many of the scientific advances that support science-based plant breeding. 'Schlegel's list' of notables over the past 200 years is fascinating. It is a fair sampling of key developers of plant breeding. I enjoyed every chapter and recommend the book to students, teachers, breeders, anthropologists, geneticists, and anyone interested in the development of historical and future aspects of agriculture."

Calvin O. Qualset, PhD
Professor Emeritus,
University of California,
Davis

NOTES FOR PROFESSIONAL LIBRARIANS
AND LIBRARY USERS

This is an original book title published by Haworth Food & Agricultural Products Press™, an imprint of The Haworth Press, Inc. Unless otherwise noted in specific chapters with attribution, materials in this book have not been previously published elsewhere in any format or language.

CONSERVATION AND PRESERVATION NOTES

All books published by The Haworth Press, Inc., and its imprints are printed on certified pH neutral, acid-free book grade paper. This paper meets the minimum requirements of American National Standard for Information Sciences-Permanence of Paper for Printed Material, ANSI Z39.48-1984.

DIGITAL OBJECT IDENTIFIER (DOI) LINKING

The Haworth Press is participating in reference linking for elements of our original books. (For more information on reference linking initiatives, please consult the CrossRef Web site at www.crossref.org.) When citing an element of this book such as a chapter, include the element's Digital Object Identifier (DOI) as the last item of the reference. A Digital Object Identifier is a persistent, authoritative, and unique identifier that a publisher assigns to each element of a book. Because of its persistence, DOIs will enable The Haworth Press and other publishers to link to the element referenced, and the link will not break over time. This will be a great resource in scholarly research.

Concise Encyclopedia
of Crop Improvement
Institutions, Persons, Theories, Methods, and Histories

HAWORTH FOOD & AGRICULTURAL PRODUCTS PRESS™
Crop Science

Biodiversity and Pest Management in Agroecosystems, Second Edition by Miguel A. Altieri and Clara I. Nicholls

Plant-Derived Antimycotics: Current Trends and Future Prospects edited by Mahendra Rai and Donatella Mares

Concise Encyclopedia of Temperate Tree Fruit edited by Tara Auxt Baugher and Suman Singha

Landscape Agroecology by Paul A. Wojtkowski

Concise Encyclopedia of Plant Pathology by P. Vidhyasekaran

Molecular Genetics and Breeding of Forest Trees edited by Sandeep Kumar and Matthias Fladung

Testing of Genetically Modified Organisms in Foods edited by Farid E. Ahmed

Fungal Disease Resistance in Plants: Biochemistry, Molecular Biology, and Genetic Engineering edited by Zamir K. Punja

Plant Functional Genomics edited by Dario Leister

Immunology in Plant Health and Its Impact on Food Safety by P. Narayanasamy

Abiotic Stresses: Plant Resistance Through Breeding and Molecular Approaches edited by M. Ashraf and P. J. C. Harris

Teaching in the Sciences: Learner-Centered Approaches edited by Catherine McLoughlin and Acram Taji

Handbook of Industrial Crops edited by V. L. Chopra and K. V. Peter

Durum Wheat Breeding: Current Approaches and Future Strategies edited by Conxita Royo, Miloudi M. Nachit, Natale Di Fonzo, José Luis Araus, Wolfgang H. Pfeiffer, and Gustavo A. Slafer

Handbook of Statistics for Teaching and Research in Plant and Crop Science by Usha Rani Palaniswamy and Kodiveri Muniyappa Palaniswamy

Handbook of Microbial Fertilizers edited by M. K. Rai

Eating and Healing: Traditional Food As Medicine edited by Andrea Pieroni and Lisa Leimar Price

Physiology of Crop Production by N. K. Fageria, V. C. Baligar, and R. B. Clark

Plant Conservation Genetics edited by Robert J. Henry

Introduction to Fruit Crops by Mark Rieger

Generations Gardening Together: Sourcebook for Intergenerational Therapeutic Horticulture by Jean M. Larson and Mary Hockenberry Meyer

Agriculture Sustainability: Principles, Processes, and Prospects by Saroja Raman

Introduction to Agroecology: Principles and Practice by Paul A. Wojtkowski

Handbook of Molecular Technologies in Crop Disease Management by P. Vidhyasekaran

Handbook of Precision Agriculture: Principles and Applications edited by Ancha Srinivasan

Dictionary of Plant Tissue Culture by Alan C. Cassells and Peter B. Gahan

Handbook of Potato Production, Improvement, and Postharvest Management edited by Jai Gopal and S. M. Paul Khurana

Carbon Sequestration in Soils of Latin America edited by Rattan Lal, Carlos C. Cerri, Martial Bernoux, Jorge Etchevers, and Eduardo Cerri

Medicinal and Aromatic Crops: Harvesting, Drying, and Processing edited by Serdar Öztekin and Milan Martinov

Mycorrhizae in Crop Production edited by Chantal Hamel and Christian Plenchette

Integrated Nutrient Management for Sustainable Crop Production edited by Milkha S. Aulakh and Cynthia A. Grant

Asian Crops and Human Dietetics by Usha Rani Palaniswamy

Concise Encyclopedia
of Crop Improvement
Institutions, Persons, Theories, Methods, and Histories

Rolf H. J. Schlegel, PhD

Haworth Food & Agricultural Products Press™
An Imprint of The Haworth Press, Inc.
New York

For more information on this book or to order, visit
http://www.haworthpress.com/store/product.asp?sku=5891

or call 1-800-HAWORTH (800-429-6784) in the United States and Canada
or (607) 722-5857 outside the United States and Canada
or contact orders@HaworthPress.com

Published by

Haworth Food & Agricultural Products Press™, an imprint of The Haworth Press, Inc., 10 Alice
Street, Binghamton, NY 13904-1580.

PUBLISHER'S NOTE
The development, preparation, and publication of this work has been undertaken with great care.
However, the Publisher, employees, editors, and agents of The Haworth Press are not responsible
for any errors contained herein or for consequences that may ensue from use of materials or
information contained in this work. The Haworth Press is committed to the dissemination of ideas
and information according to the highest standards of intellectual freedom and the free exchange of
ideas. Statements made and opinions expressed in this publication do not necessarily reflect the
views of the Publisher, Directors, management, or staff of The Haworth Press, Inc., or an
endorsement by them.

Cover design by Kerry E. Mack.

Library of Congress Cataloging-in-Publication Data

Schlegel, Rolf H. J.
    Concise encyclopedia of crop improvement : institutions, persons, theories, methods, and
histories / Rolf H. J. Schlegel.
        p. cm.
    Includes bibliographical references and index.
    ISBN: 978-1-56022-146-3 (alk. paper)
    1. Plant breeding—History. 2. Crop improvement—History. I. Title.

SB123.S324 2007
631.5'209—dc22

2006038235

Who would bring it into being that there are henceforth
growing three or four spikes, where previously only
one spike stood, he proved its homeland a service,
which is to be valued more highly than the deeds
of many kings, commanders and poets.

FREDERICK II, The Great, King of Prussia (1712-1786)

ABOUT THE AUTHOR

Rolf H. J. Schlegel, PhD, DSc, is Professor of Cytogenetics and Applied Genetics, with over thirty years of experience in research and the teaching of advanced genetics and plant breeding in Germany and Bulgaria. Professor Schlegel is the author of more than 150 research papers and other scientific contributions, co-coordinator of the international research projects, and has been a scientific consultant at the Bulgarian Academy of Agricultural Sciences for several years. He is currently working as an R&D director in a private company in Germany. His books include *Encyclopedic Dictionary of Plant Breeding and Related Subjects* (Haworth 2003), and he contributed "Rye (Secale cereale L.) - A Younger Crop Plant with Bright Future," to *Genetic Resources, Chromosome Engineering, and Crop Improvement: Cereals, Volume 2* (Eds. R. J. Singh and P. P. Jauhar).

CONTENTS

Preface

According to the World Bank, since 1960 yield growth has accounted for 92 percent of the growth of world cereal production. Genetic improvements have accounted for roughly half the yield growth of major crops. The contribution of genetic improvements to yield growth in developing countries has been similarly impressive; for example in India, yield gains in maize and rice increased by 300 percent since 1940 while sorghum, wheat, and soybean yields have doubled. It shows the great importance of *plant breeding* not only for agriculture but also for future societal development on earth.

During the past fifty years, no other branch of biology has developed so fast and as comprehensively as genetics, which analyzes the inheritance and development of microorganisms, plants, and animals, including humans. Numerous related disciplines contributed to the knowledge of the basis and structure of heritable factors, their reproduction, modification, and new entities, their intraindividual and interindividual transfer, and their permanent interaction with natural and artificial environments. Thus, modern genetics with its many branches became a fundamental discipline of biology. Genetics also influences different fields of research and stimulates new scientific approaches, including microelectronics.

Plant breeding is a field that is strongly driven by genetics, particularly during the past century. It seems to be just the beginning of a tremendous progress of targeted modification, reconstruction, and design of plants, in addition to the other challenges of breeding.

Nevertheless, plant breeding occurred independent of the development of scientific biology, philosophy, and politics. Introduction and improvement of crop plants was simply a way to (better) feed humans and animals. That is, plant breeding is a continuous process of optimization of plants for specific environments and utilization. The

Concise Encyclopedia of Crop Improvement
doi:10.1300/5891_a

19th century clearly shows that breeding activities and progress were performed without knowledge of Mendelian laws; clearly, though, breeding is also a social process. To take the most obvious sense in which this is true, there would be no professional breeders without education, societal support for the research stations, universities, and private businesses within which breeding takes place.

Because of the enormous growth in knowledge, the complexity of inheritance and breeding, and the highly specialized genetic and agricultural research, an understanding of the step-by-step (and later more systematic development) of breeding can become obliterated.

This book is a modest approach to gathering the many data, information, persons, methods, and historical developments—which are now spread across a wide range of references—into one volume. There is no similar book available on the market in English, German, French, or Spanish. The author felt that it was time to fulfill a certain need for this in the literature. This topic is usually only mentioned in introductory chapters of textbooks, papers, or college/university lessons. Moreover, this is done mostly in reference to progress in genetics, not from an original point of view.

At present, it is difficult to provide students with a historical summary in a comprehensive manner without pointing at the numerous literature references. And this becomes increasingly difficult owing to the rapid development of present plant breeding and biotechnology.

As plant breeding and adequate research becomes more and more a global task of private enterprises and several national and international organizations, the exchange of information and communication usually takes place in English, which is the predominant language in this sphere. This book can therefore be of advantage for a worldwide readership. Students of plant breeding, genetics, biotechnology, or biology might not be the only beneficiaries—breeders, teachers, or other interested persons may also benefit.

The plentitude of results, methods, and crops make it difficult for the author to consider the complete area of breeding history and to convey all knowledge based on personal experience. However, it offers the advantage of being a single representation that is more systematic than contributions authored by different authors and views. I hope this approach will prove to be successful.

Acknowledgments

I express my thanks to Mariana Atanasova, MSc, of Doubroudja Agricultural Institute; General Toshevo/Varna (Bulgaria); Mrs. W. Mühlenberg, Institute of Plant Genetics and Crop Plant Research, Gatersleben (Germany); Prof. T. Lelley, Institute of Agrobiotechnology, Tulln (Austria); Dr. K. Soon-Jong, National Institute of Agricultural Biotechnology, Suwon (Korea); Ing. Mag. M. Höller, Wintersteiger AG (Austria); Dr. B. Leithold, Institute of Plant Breeding, Martin Luther University Halle (Germany); and Drs. E. D. Budashkina, T. T. Efremova, and E. A. Salina, Institute of Genetics and Cytology, Siberian Branch, Academy of Sciences of Russia, Novosibirsk (Russia), for substantial contributions to the manuscript, proofreading, and providing several photographs.

Concise Encyclopedia of Crop Improvement
© 2007 by The Haworth Press, Inc. All rights reserved.
doi:10.1300/5891_b

User's Guide

This book provides a representative selection of information from the large amount of data on the history of plant breeding, genetics, and methods as well as institutions and persons associated with development of breeding and breeding research. A chronological representation was used in principle. However, parallel developments in terms of countries and cultures, as well as many temporal cross-references according to methods and persons, sometimes breaks the chronological order.

To limit the book's contents to strictly historical aspects, the explanation of technical terms or methods was avoided when possible, although they are often needed for understanding. These terms have been arranged alphabetically and included in the Glossary. A term when used for the first time is set in *italics*. Moreover, all scientific names are given in italics. Special designations, titles, citations, and variety names—as well as the names of some institutions—are set within quotation marks. When Greek letters were necessary in association with some words, they were translated into English. When possible, city names, institutes, and organizations are spelled in their national language. Cross-referenced terms and names are indicated by the symbol >>>. Names of scientists and/or family names are in ALL CAPS. However, when used as an adjective (e.g., Mendelian), as a part of variety names (e.g., Tschermak's Weisshafer), or in references, only the first letter is capitalized.

Descriptions of a given person within the "gallery of breeders" (Chapter 6) may be more or less extensive depending on the data available. A semicolon has been used in those listings to separate pieces of information.

Cross-references have been provided wherever necessary for demonstrating interrelationships, organizing the material in a clear manner, and economizing space. Additional notes are given at the end of the book, arranged by chapter.

Concise Encyclopedia of Crop Improvement
© 2007 by The Haworth Press, Inc. All rights reserved.
doi:10.1300/5891_c

Chapter 1

Introduction

The origin of new information in agriculture, horticulture, and plant breeding derives from two methods: empirical and experimental. The roots of empiricism derive from the efforts of Neolithic farmers, Hellenic root diggers, medieval peasants, farmers, and gardeners everywhere to obtain practical solutions to problems of crop and livestock production. The accumulated successes and improvements passed orally from parent to child, from artisan to apprentice, have become embedded in human consciousness via legend, craft secrets, and folk wisdom. This information is now stored in tales, almanacs, herbology, and histories; it has become part of our common culture.

More practices and skills were involved as improved germplasm was selected and preserved via seed and graft from harvest to harvest and from generation to generation. The sum total of these technologies makes up the traditional lore of agriculture, horticulture, and breeding. It represents a monumental achievement of our forebears.

Without knowledge of the development of a scientific discipline, it is not possible to judge recent achievements or to weigh the future opportunities. Otherwise, one could overestimate the present by a large margin. Plant breeding is no exception to this. Anyone who traces the development recognizes that the achievements of breeding *crop plants* (cf. Glossary) are based on centuries- and even millennium-old experiences—barley (Figure 1.1) is one example. The total number of plant species that are cultivated as agricultural, forest, or horticultural crops is estimated as nearly 7,000 botanical species. Nevertheless, it is estimated that only thirty species "feed the world" because the major crops are from a limited number of species. The

Concise Encyclopedia of Crop Improvement
© 2007 by The Haworth Press, Inc. All rights reserved.
doi:10.1300/5891_01

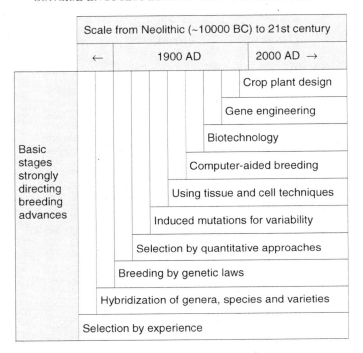

FIGURE 1.1. Major steps in plant breeding over 12,000 years of development and continuous influence by scientific achievements.

basic types of recent crops have arisen by early gathering, growing, and selection. Of course, for other crops (e.g., sugar beet, triticale, or blueberry) this step does not date back so far. Moreover, we can see how some plants actually become cultivated crops. Some fodder, ornamental, or industrial plants can be taken as examples.

Nevertheless, most crop plants are the result of a long developmental process. They derived from wild types, which are known quite well. Others are the result of spontaneous crosses, unifying the genes of two independent species within a hybrid. Under specific circumstances, the linkage of genes remains stable without a following segregation or dissociation. This is how bread wheat, domesticated plum, and rapeseed evolved.

The wild types differ from the cultivated plants not only by yield but also by many characteristics important for their existence under growing conditions untouched by humans. Therefore, wild cereals have a brittle rachis of the spike and specially shaped awns in order to dig in the soil. Seeds and fruits of wild species are typically small. The ripening fruits shatter the seeds, such as legumes, linseed, or poppy. To prevent postripening sprouting in dry areas or frost-endangered climates, some wilds form hard-shelled seeds or show dormancy, a resting condition with reduced metabolic rate found in nongerminating seeds and nongrowing buds. Some seeds (e.g., lupins) contain bitter substances to protect against damage caused by birds and other animals. Also typical are the small and irregularly shaped tubers or beets of root crops. The wild forms are substantially less demanding than the cultivated plants in their soil and climate requirements. They grow slowly and ripen less evenly. Wild species are more pubescent. This awards them the external character of a rough, resistant, and solid plant. However, they are not, in principle, more resistant than crop plants; there are susceptible wild plants and resistant crops. In summary, wild plants are adapted to produce sufficient offspring—that is, to maintain the species. A luxuriant shape usually is not required for this and is often even adverse.

In general, the crop plants are differentiated from the wild species by the missing typical wild characters—such as seed shattering, brittle spikes, bitter fruits, or branched roots—in addition to the presence of traits that are useful in agriculture, horticulture, or forestry. The resulting cultivated plant is no longer able to exist under natural environments because of such changes. Nonshattering wheat cannot sufficiently reseed. Big and delicious strawberries are eaten by animals before ripening, and fast-growing fruit trees will break as a result of weak wood and a heavy fruit load. The modern crop plant is more or less dependent on humans for its propagation by sowing, harvest, threshing, and storage. In some crops, the degree of dependence has progressed particularly far. In maize, for example, the seeds sit so tightly on the cob that self-seeding has become impossible, whereas other cereals still shatter the grains when overripe or harvested late. This shows that crop plants still can have wild features. Examples include the burst of pods in legumes or rapeseed, seed shattering in cereals, dropping of fruits in fruit trees, long germination period in parsley, bitterness in cucumbers, deep rooting, and multigerms in sugar beet.

To be classified as a cultivar, it is important that the plant shows either a majority of or the essential traits that are different from the wild characters and/or be modified for human utilization. What is still missing can be improved by subsequent breeding.

Breeding, however, does not confine itself to supplementing useful traits but also improves the existing characteristics. Through genetic engineering, even an interspecific and an intergeneric transfer of alien characters becomes feasible, in addition to the creation of novel traits (see section 4.2.2.2).

In some cases, intensive breeding has led to a situation in which cultivars lost too many of the original wild characters. Breeding toward reintrogression of wild characters is one way out of the dilemma. The enhancement of dormancy in order to reduce preharvest sprouting was such a task in rye and oat improvement.

The origin of cultivated plants is basically a process of displacing wild characters and enriching suitable traits—a process that began thousands of years ago for wheat, barley, and millet.

Chapter 2

10,000 Years of Crop Improvement

Humans have been hunters and gatherers for 99 percent of the 2 million years our species has roamed the earth. Only in the last 12,000 years have people become agriculturists. The dates of domesticated plants and animals vary by region, but most predate the 6th millennium BC and the earliest may date from 10,000 BC. According to carbon dating, wheat and barley were domesticated in the Middle East in the 8th millennium BC; millet and rice in China and southeastern Asia by 5500 BC, and squash in Mexico by about 8000 BC. Legumes found in Thessaly and Macedonia are dated as early as 6000 BC. Flax was grown and apparently woven into textiles early in the Neolithic period (4500-1800 BC). The most sweeping technology change for humans occurred in prehistory: the use of tools, the discovery of fire, and the invention of agriculture. In addition, climatic changes during the late Pleistocene—with the consequent shift of vegetation as a whole and encouragement of particular plant species to prosper and spread—have been suggested as another reason for the emergence of agriculture.

As with agriculture, plant breeding arose as a part of human development beginning in ancient times. Cultivation, the first step toward domestication, begins (at least for seed crops) with human planting of harvested seeds to provide a new crop. The harvested seeds represent a selected sample of the total variability, biased toward those characteristics of the population particularly attractive to humans and/or having the least efficient mechanisms for seed dispersal. Seeds planted by humans are, to some extent, protected from the pressure of natural

Concise Encyclopedia of Crop Improvement
© 2007 by The Haworth Press, Inc. All rights reserved.
doi:10.1300/5891_02

selection; this, together with changes in the population size, will lead to changes in variability of the crop in time.

At least six regions of domestication have been identified, including Central America, the southern Andes, the Near East, Africa (Sahel and Ethiopia), Southeast Asia, and China (>>> N. I. VAVILOV). Agriculture is thus one of the few inventions that can be traced back to several locations. From these foci, agriculture was progressively disseminated to other regions, including Europe and North America.

Actually, there are two theories on the origin of agriculture: a single origin with diffusion versus multiple independent origins. The "single origin plus diffusion" concept is articulated in a thoughtful essay by CARTER (1977). However, the presently accepted dogma is that agriculture arose as independent inventions in various parts of the world at different times. There is evidence that agriculture originated in the Middle East over 10,000 years ago. Recent studies show that Southeast Asia could stand in the same tradition.

At the old cultural centers of the world—China, Egypt, Asia Minor, around the Mediterranean Sea, Central and South America—agriculture (often with irrigation) has been practiced for thousands of years. Evidence of the importance attached to wheat, rice, soya, millet, and sorghum can be found in a regulation of the Chinese emperor CHEN-NUNG from about 2700 BC. It describes the custom of annual spring sowing, ceremonially carried out by the emperor himself.

The oldest excavation of cereals with characters of a cultivar was made in Jarmo (Kurdistan/Iraq) (HELBAEK, 1959); it was dated to the 7th millennium BC. The wild species *Triticum dicoccoides* and the cultivated type *T. dicoccum* were identified. As compared with natural conditions, even the earliest agricultural activities brought modified environments for the plants cultivated. Plant species—those that humans recognized as worthy to grow and probably used already at the time of collectors—were first classified as primary crop plants: wheat, barley, millet, rice, soybean, cotton, potato, or maize.

However, the changed conditions did not directly modify the wild types toward a cultivar. Rather, those genotypes that showed better performance, adaptation, and propagation under the culture conditions of humans became predominant among the diverse mixture of phenotypes. This is called a choice or *selection* process. It also happens in nature, but the selection contributes exclusively to the maintenance of the species and not the improvement of particular traits for human utilization. Despite the geographically diverse distribution of the domestication centers, we can identify a remarkably similar set of traits that have been selected in widely different crops. These traits jointly make up the "domestication syndrome" (HAMMER, 1984). They result from the selection of spontaneous mutations that occurred in wild populations and were selected at various stages of growth of these wild plants, for example, spikelet nonshattering in *Avena abyssinica* grown in barley fields or pod indehiscence and flat seed in the weed *Vicia sativa* grown in lentil fields. Another example is the initially rare semi-tough-rachised (domestic) phenotypes of wild einkorn *(Triticum monococcum)* that could have achieved fixation within twenty or thirty generations of selection.

The inheritance of traits has been investigated numerous times. Initially, they were analyzed as Mendelian traits because many of them display qualitative variation and discrete phenotypic segregation classes. More recently, for a limited number of crops (maize, bean, tomato, rice, and millet), these same traits have been analyzed by *quantitative trait locus* (QTL) approaches, which are more powerful because they allow a genomewide analysis of influence on several traits at the same time. It was shown that their state is often controlled by recessive alleles at one or, at most, two or three loci. Furthermore, the joint involvement of these genes accounts for most of the phenotypic variation, suggesting a high *heritability.* Finally, many of the genes are located in a limited number of linkage groups (chromosomes), and on these linkage groups they are sometimes closely, although not tightly, linked.

When humans first began to mechanize the harvest by using primitive tools, bigger spikes rather than smaller ones were preferentially selected. From the Romans, a harvest wagon is known that was pushed into a stand of cereals with the help of a donkey and two farmers (Figure 2.1). By fingerlike tools on the front side of the wagon, the spikes were pulled off and fell into a basket. This led to selection of larger spikes and the reduction of genotypes showing smaller growth habit. The consequence was a stepwise improvement of the cultivars. In addition, one can assume that some growers consciously attempted to make improvements in traits and species.

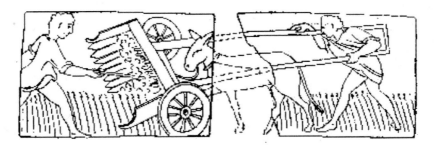

FIGURE 2.1. Reconstruction of a Gallo-Roman mow wagon (the Trevires' harvester, an ingenuous piece of agricultural machinery described notably by PLINY the younger: "In the vast fields of Gaul, large harvesters, with blades fixed at one end, are pushed on their two wheels across the crop by an ox yoked in the opposite direction. Uprooted in this way, the ears of corn fall into the harvester.") *Source:* H. SCHMALZ (1969), found in 1958 on two limestone reliefs in Buzenol and Arlon (Belgium) dated to the first half of the 3rd century AD, The Musee Luxembourgeois, Arlon.

Another factor probably led to an increase of cultivated species. Close to early settlements, some species became more enriched and showed better growth than those on distant and poorer soils. In time, the area of settlement was enriched with nutrients by animal and human excrement and other waste. Plant species that had not been recognized as worthy to collect or to grow on poor places were now recognized as suitable crop plants. Hemp, poppy, castor, cabbage, beets, mangold, carrots, parsley, and some pharmaceutical plants became such cultivated species. These plants particularly like nutrients and are called *anthropochores*. One of the latest plant species that seems to have so become a crop plant is stinging nettle *(Urtica dioica)*. There are breeding approaches for different end uses.

In this context, a third origin of cultivated plants must be mentioned. Within the fields of primary crop plants grew many weeds, for example, brittle wild rye occurred within stands of emmer wheat. As long as these weeds were poor wild types, they were eliminated during the harvest from cropped plants because of their brittleness, dispersal fruits, or small grain size. However, as soon as the accompanying weeds became adapted to the ripening time and cropping conditions of the main crop or showed other inheritable features useful for humans, they were also used as crop plants, often when the main crop suffered from drought or winter killing. Important crop plants—such as rye, oats, buckwheat, pea, lentil, white mustard, gold-of-pleasure *(Camelina sativa),* rocket salad *(Eruca sativa* syn. *E. vesicara),* and spergularia *(Spergularia maxima)*—originated within emmer wheat and barley and within emmer wheat and linseed.

In particular, the development of rye *(Secale cereale)* can be traced back over 5,000 years. With the distribution of wheat toward northern regions, rye was superior to wheat in sub-Alpine and steppe climates. Because of its cold and nutritional tolerance, it became the predominant cereal. It reached northern Europe about the 3rd or 2nd millennium BC. The deterioration of climate some 3,000-3,500 years previous promoted rye's distribution in Central Europe. Despite the poor quality of flour, it became the most important cereal crop of Celts, Teutons, and Slavs. The cultivated oat *(Avena sativa)* followed a similiar path. Prehistoric excavations seem to confirm that it came as weed within barley to Western Europe about 2,500 years ago.

The spatial scale of development of agricultural societies widely correlates with the distribution of domesticated crops. Agriculture appears to have spread at speeds averaging 1 to 5 km per year, having started around the 7th to 6th millennium BC in southeastern Europe and being completed around the 3rd millennium in northwestern Europe.

As a weed in maize and bean fields, the tomato *(Lycopersicon esculentum)* was distributed by humans from central Peru to Mexico. Only in Mexico was it cultivated, although it remained long as garden weed. The anthropochoric hemp *(Cannabis sativa)* took the way as crop plant from a weed of the Altai region of Central Asia to an accompanying plant in nomad settlements.

Crops that arose from primary crops are designated as *secondary crop plants.* Of course, new cultivated plants can develop within secondary crops. Common corn cockle *(Agrostemma githago,* Caryophyllaceae) and rye brome *(Bromus secalinus,* Gramineae) are two that show characteristics (tough rachis or grain size) that evolved under the influence of recent cultivation within rye. Within common oat *(Avena sativa)* grown in the poor soils of northwest Europe, some wild oats *(A. fatua, A. strigosa,* and *A. elatior)* achieved cultivated characters as well.

Surely, humans consciously affected genesis of cultivated plants by their preference for the most useful phenotypes, best seeds, and vegetative propagules for reseeding. We know today that the best-looking plants do not yield the best descendants, so it is not controversial that this practice sometimes led to improved cultivars and races.

2.1. THE OLD WORLD

The agriculture, horticulture, stock farming, fishery, and medicine of old Mesopotamia show extensive and versatile biological knowledge. Mesopotamian civilizations are largely based on Semitic populations that existed between the Tigris and Euphrates Rivers in what is now Iraq. The climate is winter wet and summer dry, which is particularly suitable for livestock rearing and large-scale cereal cultivation; it is the source of wild wheat and barley and also of sheep and goats. The area includes the Fertile Crescent—present-day Israel, Jordan, Lebanon, Syria, Iraq, and Iran—as well as all of Asia, southwest of Russia and the Black Sea, west of India, and Afghanistan. Spelt wheat *(Triticum spelta)*—the precursor of common hexaploid wheat *(T. aestivum)*—seems to have originated from different places of Iran and northern Caucasian regions (>>> KUCKUCK, 1959).

The agricultural history of Mesopotamia can be inferred from many sources, including cuneiform tablets and inscriptions as well as archaeological remains. The agriculture and horticulture of this region are richly annotated in biblical sources. Sumerians and Akkadians recorded long lists of plant names. In the so-called Garden Book of the Babylonian emperor MARDUK-APAL-IDELINA (8th century BC), the cultivated plants mentioned are barley, emmer and common wheat, durra millet *(Sorghum vulgare* or *S. bicolor)*, beans, peas, lentil, onions, garlic, leak, beets, radish, cucumbers, melons, sesame, linseed, the carob *(Ceratonia siliqua)*, olive, almond, and pomegranate. From the south (Egypt) came the date; from the southeast (India) came the fig, pomegranate, and citron; and from the north to northeast came the roses, lily, grapevine, apple, pear, peach, pistachio, plum, mulberry, quince, and walnut.

2.1.1. Sumeria

Sumeria was one of the advanced cultures of the 4th millennium BC, probably arising from non-Semitic populations of the East. Until recently, it was a lost culture unknown to HERODOTUS (484-424 BC). He describes exuberantly the full growth of barley and wheat: "200- to 300-fold harvests thrive (!). How beautifully the millet would thrive, it wanted conceal in order appearing not untrustworthy." He centered Sumeria in the Euphrates Valley in the Chaldean plains. It contained the ancient city of Ur, three times its capital. Sumerians were the first to develop writing (3000 BC) in the form of cuneiform script etched on soft clay tablets, which were allowed to harden into a permanent record. Sumerians introduced canals and were among the first systematic agriculturists. By 3000 BC, there were extensive irrigation systems branching out from the Euphrates River and controlled by a network of dams and canals. The main canals were lined with burned brick and the joints sealed with asphalt. At its peak, 10,000 square miles were irrigated. The legendary SARGON I, known as Sargon the Great (2334-2279 BC), founded the Akkadian-Summerian Empire. In a tale similar to that of MOSES a thousand years later, he is discovered in a reed basket:

> Akki the irrigator made me a gardener. My service as a gardener was pleasing to ISTAR and I became King. (In G. A. BARTON, *Archaeology and The Bible,* 3rd ed., American Sunday-School Union, Philadelphia, 1920, p. 310.)

2.1.2. Mesopotamia and Babylonia

There are rich literary sources for Mesopotamian agriculture. A cuneiform text from NIPPUR called *The Dialogue Between the Hoe and the Plow* is a source of agricultural information; it is dated between 1900 and 1600 BC but may well have older origins, perhaps belonging to the Ur III (~2100 to 2000 BC). In the 2nd millennium BC, the great civilization along the Euphrates known as Babylonia formed from the union of Akkadians and Sumerians, with Babylon as its capital. Historic figures include HAMMURABI (~1,750 BC) and NEBUCHADNEZZAR, who was the King of Babylon (605-562 BC). *The Code of Hammurabi* contains many laws concerning agricultural crop practices, such as irrigation and fermentation. They produced beer from barley and wheat as well as wine from grapes and dates; vinegar was a by-product.

The Hanging Gardens of Babylon, long considered one of the seven wonders of the ancient world, was supposedly built for NEBU-CHADNEZZAR's homesick bride. Spiral pumps irrigated gardens, which were 20 m high with royal chambers located under terraces. Babylonian agricultura images include a plow containing a seed drill, beer drinking, and water lifting with a shaduf. A cuneiform tablet—the restoration of a document from 1500 BC from the ancient Sumerian site of Nippur—might be the first farmer's almanac. It consists of a series of instructions addressed by a farmer to his son, guiding him throughout the year's agricultural activities. A document tablet from the same period described a myth (INANNA and SHUKALLI-TUDA: "The Gardener's Mortal Sin") that reveals the agricultural and horticultural techniques of windbreaks—planting shade trees in a garden or grove to protect plants from wind and sun. A cuneiform tablet from about 1300 BC shows a map of fields and irrigation canals. An Assyrian herbalist in the 7th century BC named 900 to 1,000 plants. Examination of clay tablets in the library of King ASSUR-BANIPAL of Assyria (668-626 BC) identified 250 vegetable drugs, including asafetida, calamus, cannabis, castor, crocus, galbanum, glycyrhiza, hellebore, mandragon, mentha, myrrh, opium, pine turpentine, styrax, and thymus. Their special cultivation and selection can be assumed.

Sumerians and Babylonians knew the sexual *dimorphism* and *dioecious* habit of the date palms. They artificially pollinated the female flowers either by hand (Photo 2.1) or by hanging the male flower bunch in the crowns of the fertile trees. The number of male trees could be kept small, and the fruit setting of the females could be increased. Planting of palm trees distributes about one male plant among 100 females (SWINGLE, 1913). In the history of breeding, this is the first documented evidence for guided *allogamy* by humans, opening the chance for breeding new varieties and broadening the genotypic variability for selection. ROBERTS (1929) reported that, in four Sahara oases, more than 400 varieties of date palms could be distinguished by size, shape, and taste of fruits. In Mesopotamia and Egypt, which

PHOTO 2.1. Assyrian relief of the period of emperor ASSURNASIRAPAL II (883-859 BC), illustrating pollination of female flowers of date palms by winged genie wearing bird masks. *Source:* Institut f. Pflanzengenetik u. Kulturpflanzenforschung, Gatersleben. Used with permission.

were better known for their breeding and handling, no other dioecious species was grown at that time.

Possibly maize is another example of early breeding-like approaches, although no record of this is available. However, primitive varieties of maize were found in a South American bat cave dated back to 4000 BC (see the following text).

2.1.3. Judea

Much is known about the culture of Judea (1200-587 BC) because many books of the Bible have come to us almost intact. In the biblical literature, common agricultural and horticultural practices are discussed, but the interpretations are usually religious or moral. Still, a reading of the scriptures tells us much about the agriculture of this period. The basic roots of these "desert" people are well represented, although in these early times the annual rainfall was higher and soils more fertile than nowadays. Plants, plant products, and agricultural technology are referred to in hundreds of verses. Genetics can even be found in the Old Testament of the Bible. In Leviticus 19:19, the Israelites are warned not to let their cattle breed randomly and not to sow their fields with mingled (contaminated) seed. About breeding activities, it is written further:

> Yet I had planted thee a noble vine, wholly a right seed: how then art thou turned into the degenerate plant of a strange vine unto me? (JEREMIAH 2:21)

> I am the true vine, and my father is the husbandman. Every branch in me that beareth not fruit he taketh away: and every branch that beareth fruit, he purgeth it, that it may bring forth more fruit . . . As the branch cannot bear fruit in it, except it abide in the vine; no more can ye, except ye abide in me. I am the vine, ye are the branches: He that abideth in me and I in him, the same bringeth forth much fruit: for without me ye can do nothing. If a man abide not in me, he is cast forth as a branch, and is withered; and men gather them and cast them into the fire, and they are burned. (JOHN 15:1-6)

In a related context, it is thought that fruits of wild figs were artificially ripened. The fruit depends on pollination by wasps *(Ceratosolin arabicus).* When sycamore (fig) was introduced to Egypt, apparently the wasp was not introduced and seeds were not produced. To ripen the fruit without pollination, the ancient system was to "scrape" the fruit with iron claws as THEOPHRASTOS of Eresus, city of Lesbos (372-287 BC), mentioned. This practice is still carried out in Egypt and Cyprus. The wounding acts to increase ethylene, which induces ripening. Ethylene is the most recent addition to the list of plant hormones. 2(chloroethyl) phosphoric acid, an ethylene-generating substance, is now commercially used to induce ripening as well as latex flow in rubber.

2.1.4. Egypt

Egyptian culture dates back to the dawn of civilization, and remnants exist in a continuous 6,000-year-old record. The artistic genius engendered by Egyptian civilization and the dry climate there has made it possible to reconstruct a history of agricultural technology. Ancient Egypt is shown to be the source of much of the agricultural technology of the Occident. From 4000 to 3000 BC these mingled peoples of the Nile valley formed a government, constructed the first pyramids, and established highly advanced agricultural technology. The ancient names for Egypt underscore the relation between the land, the people, and its agriculture. These include *"Ta-meri,"* the beloved land cultivated by the hoe; *"Ta Akht,"* the land of flood and fertile soil; *"Kmt,"* the black soil; *"Tamhi,"* the land of the flax plant; and *"Nht,"* the land of the sycamore fig tree.

The ancient Egyptian god of vegetation, OSIRIS, is credited with introducing the skills of agriculture to the Egyptians. He became the god of the dead and the underworld, following his slaying by his brother SET and restoration to life by his wife and sister ISIS. OSIRIS is sometimes depicted with green skin in paintings or in statues made from green stone, reflecting his aspects of agriculture, vegetation, fertility, and resurrection. He was recorded as the first to make mankind abandon cannibalism and is credited with introducing culture of the vine and fermentation of its fruit to produce wine. This may be the oldest account of a biochemical process. The legend of OSIRIS and ISIS dates back to at least 2400 BC, as recounted by the Greek historian PLUTARCH (46-120 AD). The historian Diodorus SICULUS (~1st century BC) included OSIRIS among those who had been men and were immortalized by virtue of their sagacity and good works. The story may be derived from the Sumerian goddess of fertility, ISHTAR, who could grant crops to her devotees.

Knowledge of the crops of ancient Egypt can be deduced from the artistic record, but definite proof comes from the desiccated remains of plants themselves. The chief ancient grain crops, used for bread and beer, were barley and various wheats including the diploid einkorn, the tetraploid emmer and durum wheats, and the hexaploid spelt and bread wheats. One of the ancient cereals[1] of Egypt—now marketed as *"kamut"*—has recently been introduced in the United States and Europe.

The vegetables of ancient Egypt included a number of root crops, leafy salad crops, legumes, and various cucurbits. The ancient root crops, such as the pungent alliums (garlic, *Allium sativum* and onion, *A. cepa*) and the radish *(Raphanus sativum),* continue to be popular in modern Egypt. Among the leafy salad crops were lettuce *(Lactuca sativa)* and parsley *(Petroselinum crispum).* There were a number of pulses, such as cowpea *(Vigna unguiculata)*, broad bean *(Vicia faba)*, chickpea *(Cicer arietinum)*, and lentils *(Lens culinaris)*. The cucurbits included cucumber *(Cucurbita sativa)*, melons *(Cucumis melo)*, gourds *(Lagenaria* spp.), and later watermelon *(Citrullus lanatus)*. The date and doum palm *(Hyphaene thebaica)*, also known as the gingerbread palm, and the sycamore fig *(Ficus sycomorus)* are considered predynastic Egyptian fruits although the sycamore is indigenous to east and northeast Africa. Jujube, common fig, and grape were known since the Old Kingdom; carob and pomegranate were introduced in the Middle Kingdom; olive and apple appear in the new Kingdom, and peach and pear date to the Graeco-Roman period.

About 2,000 species of flowering and aromatic plants have been found in tombs. An exquisite bas-relief depicting a visual representation of the fragrance from essential oils being extracted from an herb is found on the walls of the PHILAE Temple. Herb and spice plants—important for culinary, cosmetic, medical, and religious uses—were continually introduced. Pharaohs were horticulturally sophisticated and were collectors. From foreign campaigns they brought back exotic trees and plants to be grown in their palace or temple gardens. Queen HATSHEPSUT organized a plant expedition delivering living myrrh trees from Punt (Northeastern Africa) for the terraced gardens of her Temple at Deir el-Bahri in 1500 BC, and she kept records about the plants discovered. It might have been the first expedition to collect plant genetic diversity and to preserve them like gene bank material. Herbs, spices, aromatics, and medical plants included *"ami"* or Ethiopian cumin *(Carum copticum)*, anise *(Pimpinella asisum)*, caper *(Capparis spinosa)*, coriander *(Coriandrum sativum)*, cumin *(Cuminum cyminum)*, dill *(Anethum graveolens)*, fennel *(Foeniculum vulgare)*, fenugreek *(Trigonella foenum graecum)*, marjoram *(Origanum majorana)*, mint *(Mentha spicata, M. sativa)*, mustard *(Sinapis alba)*, rosemary *(Rosmarinus officinalis)*, safflower *(Carthamus tinctorius)*, and thyme *(Thymus acinos)*. The queen's nephew THOTHMES III (1,450 BC) had the curious plants brought from Syria carved on the walls of the Temple of Amen in Karnak, among which iris can be recognized.

Industrial and fiber crops were important in ancient Egypt. Oil of the castor bean *(Ricinus communis)* was used for illumination and as a medical plant, flax *(Linum usitatissimum)* for linen and oil, henna *(Lawsonia inermis)* for dye, papyrus *(Cyperus papyrus)* for paper, aquatic lotus *(Nymphaea lotus* and *N. caerulea)* for seed and decoration, and acacias *(Acacia* spp.) for gums and oils.

The ancient Egyptians during the period of RAMSES III (1198-1166 BC) loved flowers, as evidenced by murals portraying court ladies wearing Egyptian lotus blooms, container-grown plants, and funeral garlands. They founded gardens with wide places for walking and with all kinds of sweet fruit trees or flowers.

Small shrubs were grown in large earthenware pots, the forerunners of the present day potted plant industry. Plant dyes were an important part of the cosmetic arts. Aromatic ingredients from flowers were incorporated into oils and fats for use in perfumes, unguents, and ointments. The basic tools of agriculture—the ax, the hoe, and the plow—are independent Egyptian inventions. The prototype hoe can be seen as a modification of a forked branch, while the more developed form has a hafted wooden blade. The plow was at first a modification of the hoe and was originally drawn through the ground, perhaps first by a man using a rope but later, in the Old Kingdom, by a pair of oxen. Later, metal plowshares were added. In the New Kingdom, handles were lashed by ladder-like cross-pieces and the shaft was bound to a double yoke over the oxen's horns. Sowing followed plowing. Often the sowers scattered seed in front of the plow so that the oxen treaded it in, whereas fine seed as flax was shaken directly into the furrows. If the seed was sown after flooding then sheep, goats, or swine were driven to tread in the seed. HERODOTUS described the seeding technology as follows:

> . . . for they have not the toil of breaking up the furrow with the plow, nor of hoeing, nor of any other work which all other men must labor at to obtain a crop of corn; but when the river has come of its own accord and irrigated their fields, and having irrigated them has subsided, then each man sows his land and turns his swine into it; and when the seed has been trodden into by the swine he waits for harvest time. (In W. DURANT, *Our Oriental Heritage,* Simon & Schuster, New York, 1954, p. 156.)

Harvest and postharvest handling of grain were favorite themes in Egyptian art. Early sickles used to cut wheat had flint teeth set in a wooden or bone haft followed by curved sickles with a short handgrip. Metal sickles were common in the New Kingdom. Wheat was bound into sheaves and loaded onto donkeys for transport to storage. Fruit was collected and packed in shallow baskets, and was artfully arranged. Evidence of grain storage in buried baskets or earthenware jars dates to Neolithic times. Later, the storage of grain and other provisions became a state function, and communal silos and granaries were constructed. In the temple of Abu Simbel, built by King RAMSES II (XVIII dynasty), the following words are carved:

> I [the God PTAH] give to thee [RAMSES II] constant harvests, to feed the two lands at all times; the sheaves thereof are like the sand of the shore, their granaries approach heaven, and their grain-heaps are like mountains. (In W. J. DARBY, P. GHA-LIOUNGUI, P. L. GRIVETTI, *Food: The Gift of Osiris,* Vol. 2, Academic Press, London, 1977, p. 167.)

The Roman world considered Egypt the "breadbasket of the world." Grapes were much appreciated in ancient Egypt, as evidenced by abundant depictions of grapes, grape culture, and wine making. C. PLINIUS the Elder (23-79 AD) reported that vines were grown directly on the field surfaces, but there are many representations of trellises and arbors. The round arbor was a common feature between the New Kingdom and the Graeco-Roman period. Greek authors confirmed that various cultivars of *Vitis vinifera* were developed in Egypt; for example, the variety *"Gutedel"* (in German; in French, *"Chas-selas"*) is still well distributed in Europe. The latter is characterized by red-veined and five-lopped leaves found on 4,000-year-old reliefs and walls at Luxor. The Phoenicians brought it from the region of Lebanon to the Rhone valley of France and later further north.

The beginnings of *biotechnology* (see Chapter 4) are directly traced to the manufacture of bread and wine. The harvest, threshing, grinding of grain to flour, and subsequent sieving are abundantly illustrated. Fermentation by the use of leaven, a mass of yeast, was a development that changed the making of bread. By 1200 BC, over thirty different forms of bread and cakes were mentioned.

Alcoholic fermentation was carried out in pots with bread or flour to make beer, or with sugary fruit juices (particularly of grape and of dates and pomegranates) to make wine, or with honey to make mead. Wine is specified as early as the first dynasty and was associated with

HORUS, the falcon-headed god, son of ISIS, the Great Mother, loyal sister and wife of OSIRIS, god of the beneficent Nile. About 1200-1500 BC, the complete wine-making process is illustrated, beginning with the grape harvest from arbors, the treading by workers who maintained their balance by holding hanging cords attached to a frame, and the squeezing of the sediment collected in cloth bags, with fermentation carried out in amphorae. After fermentation, wine vessels were sealed with plugs of straw and clay designed to prevent bursting due to gas accumulation and were then impressed with official stamps containing the year of the king's rule, the district, the town, and the name of the wine. By the Graeco-Roman periods, there were literally hundreds of wine types from grapes, suggesting intense genetic selection.

2.2. PLANT CULTIVATION IN ASIA
SINCE NEOLITHIC TIMES

Asia is one of the first continents in which civilization developed and where humans learned to live by cultivating plants. In fact, more than half of the world's edible food crops originated in Asia. The prehistoric discovery that certain plants are edible or have curative powers and others are inedible or cause harm is the origin of the healing professions and its practitioners and also of the plant sciences: botany, agriculture, and horticulture. Even though Asian agricultural technology had (and continues to have) an enormous impact on the world, its agricultural history seemed mysterious to the West and they ignored it. For example, Sogdian tribes in Central Asia—the predecessors of Tadshikes—3,000 years ago bred apricot trees whose fruits had a sugar content of 70 percent. The fruits dried on the tree without dropping down.

The huge continent is often lumped with Europe and the Middle East as Eurasia, but there is no precise demarcation to this large landmass. However, eastern Asia, which was known to the West as the Orient, is ecologically separated from the western part of the continent by the formidable boundaries of deserts and massive mountain ranges. Ancient contacts between east and west was largely indirect, through trade by sea via India and Arabia and overland through the Silk Road.

The world's traditional agronomic and horticultural crops must be viewed as an introduction from Asia. In contrast, as a result of glaciation during the Ice Age, the northern areas of the world (e.g., Siberia, Northern Europe, Canada, and the continental United States) contain few native species and have contributed relatively few world crops: pasture grasses, sunflower,[2] and some small fruits such as strawberry, blueberry, cranberry, and lingonberry. Consequently, Asian germplasm resources must be given special consideration for future crop improvement.

SAUER (1969) has long speculated that the origin of plant domestication occurred in southeastern Asia, but later archeological evidence for agriculture indicates the Middle East. The major reason for this is that the oldest civilizations about which we have much information emerged in the Fertile Crescent. REED (1977), based on evidence of crop remains in the Spirit Cave in Thailand, has pushed the evidence for agricultural beginnings as early as 10,000 to 14,000 years ago. There is now near consensus that plant and animal domestication emerged in China at least 7,000 to 9,000 years ago. Of course, even with an independent origin, diffusion of information and germplasm strongly influenced agricultural development.

The earliest evidence of agriculture is found in northern China from 5000 to 7000 BC, although an excavation in South Korea in 2001 dates 59 cultivated grains of rice over 14,000 years. The P'EI-LO-KANG and related cultures occupied the loesslands, and they domesticated foxtail millet *(Setaria italica)* and panic millet *(Panicum miliaceum)*. Large Neolithic villages (YANG-SHAO culture) have been uncovered, yielding pottery, hoes, polished axes, and weights for digging sticks. Neolithic cultures flourished from the Korean Peninsula to Manchuria to Vietnam. Artifacts include beautiful painted pottery as well as storage jars with seed. Crops include bamboo for shoots, persimmon, grass seed, walnut, pine nut, chestnut, and mulberry.

By 4000 BC, there were large farming villages. Evidence of rice cultivation appeared in the lower Yangtze Valley. Chinese civilization was based on rice and millet. *Brassica* seeds were found in pots, and hemp was grown as an edible seed and as a fiber plant for clothing. The mulberry and silkworm culture began then—the first known domestication of insects, predating honeybee culture by thousands of years.

Agriculture spread to Manchuria by 3500 BC. Millets were predominant. Rice was farmed from Taiwan to central India before 2500 BC. Melons, sesame, and broad bean were also cultivated. Wheat and barley were introduced from Afghanistan.

The founder of Chinese agriculture and medical botany is considered to be the mythical emperor SHEN-NUNG (2737-2697 BC). In the 1st millennium he was known as the "Divine Cultivator" of five grains, inventor of the plow and soil tests for suitable crops, and the originator of ceremonials associated with sowing vegetables and grains.

CONFUCIUS lived from 551 to 470 BC. He is considered the author of *Book of Songs*. It includes 300 traditional songs of the ZHOU Dynasty (1066-221 BC) with hundreds of references to food, providing an agricultural picture of the age. Mention is made of clearing artemisia, thistles, and weeds. Fiber crops were silk, hemp, and kudzu. The staple food was millet. *Book of Songs* mentions 44 food plants (as compared with the Bible, which names 29). These include:

> *Grains*: millets *(Panicum and Setaria)*, barley, and rice.
>
> *Vegetables*: kudzu *(Pueraria lobata)*, hemp, Chinese cabbage *(Brassica* species), Chinese chives *(Allium tuberosum)*, daylily *(Hemerocallis flava)*, bottle gourd *(Lagenaria)*, melon *(Cucumis melo)*, soybean *(Glycine max)*, lotus *(Nelumbo* species), yarro *(Achillea siberica)*, mugwort *(Artemisia)*, motherwort *(Leonurus species)*, mallow *(Malva verticillata)*, huanlan *(Metaplexis stauntoni)*, plantain *(Plantago major)*, poke *(Phytolacca acinosa)*, huo *(Rhynchosia* species), cocklebur *(Xanthium strumarium)*, sweet flag *(Acorus gramineus)*, water plantain *(Alisma plantageo)*, water fern *(Marsilea quandrifolia)*, elm *(Ulmus)*, and bamboo *(Bambusa* ssp.).
>
> *Fruits and nuts*: peach *(Prunus persica)*, pear *(Pyrus* species), plum *(P. salicina)*, Japanese apricot *(P. mume)*, Chinese jujube *(Ziziphus jujuba* and *Z. spinosa)*, raisin tree *(Hovenia dulcis)*, Chinese chestnut *(Castanea mollissima)*, white mulberry *(Morus alba)*, oak *(Quercus)*, pine *(Pinus)*, brown pepper *(Zanthoxylum piperitum)*, and Japanese quince *(Chaenomeles japonica)*.

From the 3rd to the 2nd century BC, there is increasing complexity in culture. A machine to winnow grain was invented. A machine controlled the separation of chaff and grain, and air movement was generated by a fan. Various types of wheelbarrows, unknown in the West until the 11th century AD, appeared in China in the 1st century BC. At this time, seed was sown by hand along ridges, with extensive hoeing to destroy weeds and to create a soil mulch for conserving moisture. A multitube seed drill dates to this period.

2.2.1. China

In the 1st century, FAN SHENG-CHIH's *Agricultural Manual* describes multiple cropping (i.e., winter wheat or barley followed by millet). Pretreatment of seed—for example, steeping in fertilizer made from cooked bones, manure, or silkworm debris, to which aconite or other plant poisons were added—was known. Irrigation of rice by water trapping for dry land fields in the north was practiced. Cultivation in pots and pot irrigation can be found. Ridge cultivation, scheduling of fertilization, watering and planting, organic matter recycling, and soil adaptation to crops were described and/or can be reconstructed. Around 1100 BC, during the SHANG dynasty (1700-1027 BC), SHANG emperors commanded the breeding of early varieties of rice based on accessions of Burma (100 instead of 180 days for vegetation period).

Crops and foods mentioned from the HAN dynasty (206 BC-220 AD) include wheat, barley, glutinous and spiked millets, soybeans, rice, hemp, and *Vigna;* gourds, taro, mulberries, Artemisia, melons, scallions, perilla, sesame, and elm (leaves and seeds eaten), mustard greens (Chinese cabbage), mallow (*Malva* spp.), leeks, onions, water peppers (an aquatic green similar to watercress), and some unidentified herbs are also mentioned. Other foods include lotus, longan, litchi, cinnamon, fagara or Chinese pepper *(Zanthoxylum),* magnolia buds, peonies, rush shoots, galangal, daylilies, true oranges, grape, chestnuts, water caltrop *(Trapa bicornis),* bamboo shoots, sugarcane, honey, assorted wild herbs, and wild ginger. Small beans appear to be adzuki bean or red bean.

Peasant agriculture was based on small, walled communities with contiguous fields. Plows were made of wood, stone, and then iron, but the basic tool was the hoe. Soil was amended with natural fertilizer. Cereals were chiefly rice and millet with some wheat and barley. The major beverage was tea, first used as a medicinal but later used as a ubiquitous beverage by the TANG dynasty (618-907 AD). Vegetables were widely consumed—including soybean and its sprouts, garlic, onion, and many crucifers—as well as berries and fruits including peach and Asian pears.

The agricultural development of China after the HAN period was influenced by the influx of new crops from Central Asia and India, such as tea. During the TANG dynasty, the "Golden Age," crops such as spinach, sugar beet, lettuce, almond, and fig entered China from Central Asia and the Middle East, while palm sugar (jaggery), date, yam *(Dioscorea)*, new types of rice, taro, myrobalan plum, citrus, cassia, banana, *Canarium*, and litchi entered from the south. Distillation, a Chinese invention, appeared and spread to Europe through Arab cultures. In the SONG dynasty (960-1269 AD), food production became rational and scientific. New crop cultivars were introduced, including short season rice *("Champa")* and green lentils (mungbean) from India. Watermelons and sorghum became mainstays of north China's dry landscape. Cotton and sugarcane from Central Asia was widely grown and used.

At the end of the MING dynasty (1368-1644 AD), New World crops became well known throughout China. Maize, sweet potato, potato, tomato, and peanut profoundly affected Chinese agriculture. China's food became stabilized by the end of the MING dynasty. Rice made up approximately 70 percent of grain cultivated; wheat and millet constituted the remainder.

By the QING period (1644-1912 AD), intensive agriculture produced food in surplus. Root crops such as sweet potato and potato became increasingly abundant. Maize and tomato cultivation increased. Indigenous Chinese species—particularly ornamentals, such as the tea rose, and food species such as Chinese gooseberry, kiwi, and Asian plums—entered the West.

Horticulture became embedded in the culture of China and Japan through the establishment of rural retreats and urban gardens. Cultivating flowers was considered one of the seven arts and assumed mystic importance. The peach blossom became the emblem of spring, the lotus of summer, chrysanthemum of autumn, and the narcissus of winter. Plum blossoms symbolized beauty and bamboo stood for long life. Lotus and peony were especially prized. By the 11th century AD, there were 39 cultivars of tree peony and 35 cultivars of chrysanthemum.

2.2.2. India

The origin of the subcontinent is based on an ancient collision of a landmass with Asia; this produced the Himalayas, a natural boundary between China and India, especially for the distribution of plant species. The climate varies from the wet monsoons to deserts, from the snows of the Himalayas to the hot jungles of the south. India never was a single nation. Some of the original invaders were called Aryans, who derive from peoples from the shores of the Caspian Sea, whose western branch populated Europe. This can be traced back by etymological similarities (Indo-European languages).

The Indus valley is considered to be one of the areas in which agriculture originated. Beginnings appear in the 6th millennium BC. The remains of ancient cities rivaling those of Sumer, Babylonia, and Egypt are found in two sites, Mohaenjo-daro and Harappa, which date to about 3000 BC. They include pottery, coins, seals that bear undecipherable inscriptions, and a huge granary on a brick base with an air-drying system. The original peoples of southern India were agriculturists who tilled the soil and raised crops of barley, wheat (rice was a later introduction), pea, cotton, and (occasionally) sesame and mustard.

From these beginnings, cultivation in India extended to rice, pulses, millet, vegetables, and fruits. Dams for irrigation were constructed as early as the 1st century AD. An Indian cuisine developed that is rich in spices: curry, ginger, cloves, and cinnamon. One of the early crops was cotton, known to the ancients as tree-grown wool. Evidence of herbal use in India predates the 1st millennium BC, with listings of plants for therapeutic value dating to 500 BC.

Trade between Asia and the Near East is very ancient. Evidence of silk strands in an Egyptian mummy in the 10th century BC suggests early exchanges of goods between China and Egypt, probably from overland routes that included Persia. The ancient spice trade between the Far East and Egypt is known from the biblical story of Joseph in

Genesis and from repeated allusion to spices, such as cinnamon and cassia. These valuable spices, which originated in the Maluccas or Spice Islands, reached the Near East overland or by sea via India.

The Silk Road was undoubtedly the route of Asian crops, such as citrus and peach, to the West. In fact, the peach (Latin: *persica*), domesticated in China 3300-2500 BC (FAUST and TIMON, 1995), is so named because it sojourned in Persia on the way and was mistakenly thought to have originated there. It was imported to Rome during the 2nd century BC.

Monks in the 6th century AD are reported to have brought back mulberry leaves and caterpillars to introduce the silk industry to the West. Sugarcane drifted to the West via the Arabs, an exchange that was to have an enormous effect on the history of the New World through development of the plantation system based on the labor of African slaves.

The movement of crops was not one way. During the Age of Exploration in the 16th century AD, colonial excursions of the West to the East were pursued by Portugal, which established trading centers in India due to the voyage of Vasco da Gama (1429-1624 AD). The Spanish and Portuguese introduced New World crops, which spread rapidly, mostly from Manila. Sweet potato, introduced in the latter half of the 16th century AD, was the most important introduction; it was well known by AD 1594 as *"camotl,"* the Aztec name, or by Chinese names, such as *"calle chin-shu"* (golden tuber), *"pai-shu"* (white tuber), *"fan-shu"* (Barbarian tuber), or *"kan-shu"* (sweet tuber). Peanut is first mentioned about AD 1538, and maize in AD 1555. Other New World crops included tobacco, tomato, guava, papaya, dicama, and yam bean. Throughout the centuries there has been a continuous exchange of plant material, germplasm, and agricultural knowledge.

2.3. CROPPING PLANTS IN ANCIENT AMERICA

Three great civilizations are known to have existed in ancient America—Aztec, Maya, and Inca. These were monumental cultures—similar in many respects to the Egyptian civilization of 2000 BC—with enormous temples in the form of pyramids, pictorial writing, a system of cities and government, and a developed agriculture. Archaeological evidence indicates that the New and Old Worlds are connected—through the Bering Strait—and that the first peoples of the New World migrated more than 50,000 years ago. Modern molecular studies of human DNA variability has revealed that South America was settled by Indian tribes.

Aztec is derived from *"Aztlan"* (white land), probably northwest Mexico, where by tradition the tribe originated; it is also known as *"Tenocha."* It gave the name to Tenochtitlan, a city founded by the Aztecs on an island in Lake Texcoco in the Valley of Mexico (now Mexico City). The Aztecs arrived late in Mexico, by about 1168 AD. They were the heirs to previous cultures. Aztec herbals described hallucinogenic plants, such as peyote. The life revolved around the cultivation of maize (called *"milpa"*). Milpa culture remained unchanged for 3,000 years. The maize was planted in March in holes 10 to 12 cm deep. No fertilizer was used except human feces. The Incas had bird manure (guano) and limited irrigation.

Sunflower[2] was the other common crop among American Indian tribes throughout North America. Evidence suggests that Indians in present day Arizona and New Mexico have cultivated the plant since about 3000 BC. Some archaeologists suggest that sunflower may have been domesticated before maize. The sunflower was used in many ways throughout the various American Indian tribes. The seed was ground or pounded into flour for cakes, mush, or bread. Some tribes mixed the meal with other vegetables, such as beans, squash, and maize. The seed was also cracked and eaten as a snack, as still happens in eastern Europe. There are references of squeezing the oil from the seed and using the oil in making bread. Nonfood uses include purple dye for textiles, body painting, and other decorations. The dried stalk was used as a building material. This exotic North American plant was taken to Europe by Spanish explorers some time around 1500 AD. The plant became widespread throughout present-day western Europe mainly as an ornamental, but some medicinal uses were developed. Sunflower became popular as a cultivated plant in the 18th century and more recently as an industrial crop.

In temperate zones of America, beans and squash were sowed in the same hole—the maize acted as support for the climbing beans. Maize and beans complement each other to form a complete staple food: maize alone causes deficiencies in the essential amino acid lysine, and beans alone are deficient in the sulfur-containing amino acids, such as cysteine and methionine. The mixture of beans and tortillas make a complete protein food. Beans and maize also complement agriculturally, with maize providing a support for beans and beans, as N-fixing plants, providing added nitrogen to the soil. The Aztecs had little irrigation technology and thus depended on rain. Other plants, such as sweet potato (warmer valleys), tomato, chili pepper, amaranth, pineapple, avocado, chicle-zapote, and chocolate (cacao), were grown in gardens and small fields. The "chinampas" system[3] predated the Aztecs and became the basis of their agriculture. Maize produced several crops a year. Before planting, farmers scooped rich mud from the bottom of the lakes, loaded it in a canoe, and so fertilized the "floating" fields. Seed nurseries were near the end of the canals. Today, chinampas yield seven crops per plot per year. Two are maize; others may be beans, chili, tomato, and amaranth. There were elaborate gardens and ceremonies.

Maize was also the basis of Mayan civilization. Each person was allotted plots (land was communal property), which was weeded by hand. Stalks were bent at harvest to deter birds. Grain was preserved in storage bins and underground granaries, and water was provided from reservoirs and wells. The yield was deduced from present day statistics for subsistence agriculture: ~70 kg/ha. Other Mayan crops were Tepary beans *(Phaseolus acutifolius)*, squash *(Cucurbita moschata, C. mixta)*, pumpkin, chili pepper, sweet potato, sweet cassava, chicham, papaya, avocado, achiote (as food color), gourds for bowls, hemp for fiber, sapodilla (for adhesives), copal, Brazil wood, palms, and cacao. The best varieties of cotton in recent days of the United States and Russia derived from Mexican farmers and their Mayan ancestors.

The basis of Inca society in Peru was the farmer-soldier (as in antique Rome). First fruits were given as a religious offering to the local shrine *("huaca")*. The seasonal working of Inca farmlands is illustrated in a series of monthly drawings by a Peruvian of Indian-Spanish descent in 1580 AD. Many of the local crops can be identified in pottery. Staples include potato and maize. Peru is the center of origin for potato. Tubers were preserved to a freeze-dried product *("chuñu")* by continued freezing combined with squeezing the tubers by walking on them.

The Incas called maize *"sara."* The many types included sweet maize *("choclo")* or parching maize *("kollo sara")*. Maize showing very large kernels (three to four times bigger than common ones) was grown until recent days in the highlands of Peru; it belongs to the group of *"cuzco"* maize. Crop remains of excavations documented that the Inca farmers grew this variety more than 1,500 years ago.

Quinoa, a chenopod, is a spinach-like plant whose dry seed is consumed as a grain—as is amaranth, which is still consumed in South America. It is now being reconsidered in the United States and Europe and may have industrial prospects because of its unique starch properties. Many other fruits and vegetables were cropped on terraces.

2.4. THE GREEK AND ROMAN WORLD

There is evidence of civilization appearing during the Bronze Age (3200 to 2000 BC) in mainland Greece and neighboring islands. The period between 2000 and 1600 BC marks the arrival of the first Greek-speaking Indo-European populations. By this time, a complex urban civilization had developed in Crete. The Mycenaean kingdoms (by 1600 BC) developed an agriculture including irrigation and the draining of Lake Coapis. The culture reached its high point in the 5th century BC.

However, before the time of ARISTOTELES there must have been a rich special literature on agriculture and plant breeding. By name are known the authors ARCHYTAS, ANDROTION, APOLLODORUS of Lemnos, CHARES of Paros, HIPPON, KLEIDEMOS, and LEOP-HANES. The latter can be associated with the 4th millennium BC. The written fragments point to collections of empiric knowledge. Such fragments are known from KLEIDEMOS, who for the first time mentioned the problem of plant diseases.

The time frame between 750 and 450 BC is known as the era of Hellenism. This was the great period of Greek colonization, probably instigated by the shortage of arable land on the Greek mainland. Colonialization extended in the Mediterranean region as far west as Spain and as far east as the northern boundary of the Black Sea. This period coincides with innovations in all fields of thought and technology.

The basic techniques of agriculture and horticulture were well established in the ancient cultures (i.e., 3000 BC-500 AD). The contribution of this period to progress was the creation a written record of agricultural achievements. It also contained the seeds for the beginnings of what one can call scientific studies. The agricultural and horticultural achievements include the following:

- Planting and cultivation technology involving plowing, seed bed preparation, and planting
- Irrigation technology including water storage in dams and ponds, channeling of water above and below ground, water-lifting technology
- Basic technology of storing agricultural products in granaries, underground storage, and cave storages
- Fertilization and crop rotations
- Basic propagation technology, such as seed handling, grafting, layerage, and cuttage
- Improvement of crops by selection and clonal propagation
- Basic development of food technology, such as fermentation (bread and wine), drying, or pickling
- Development of gardens and parks

Greek culture, based on the domination of ideas rather than technology per se, spread throughout the entire Mediterranean basis and had a powerful influence on Roman culture. Today Greek and Latin are the basis of scientific English in botany, biology, and agriculture.

The culture of the developing West is based on a fusion of Greek culture, Babylonian and Egyptian science, and Semitic religion. The art is typified by idyllic realism, included depictions of gods, animals (particularly horses), and plants, and it includes agricultural practices. DEMOCRITUS of Abdera (460-370 BC), the founder of the atomic theory, also had theories on the nature of plants. For example, he thought plant diversity was due to differences in the atoms of which they were composed. HIPPOCRATES of Kos (460-377 BC), considered the originator of a Greek school of healing, was the first to clearly expound the concept that diseases had natural causes. The use of drugs was not ignored, and between 200 and 400 BC he mentioned the medicinal use of herbs. THEOPHRASTOS is claimed to be the founder of the botanical sciences and is thus known as the Father of

Botany. He was a student of ARISTOTELES (384-322 BC). Several books by THEOPHRASTOS remain, for example, *Historia Plantarum* and *De Causis Plantarum*. In continuation of the biological work of his teacher ARISTOTELES, he described and classified different plant species. Numerous terms that he introduced—for example, *"carpos"* (the fruit) and *"pericarpion"* (the mature ovary wall)— were in use until recent times. He taught that higher plants are sexually propagating. However, this knowledge was lost during late antiquity until CAMERARIUS again demonstrated the existence of different sexes in plants (Chapter 2.7). ARISTOTELES' description about the diversity of crop plants is particularly remarkable. In THEO-PHRASTOS' *Historia Plantarum* we read as follows:

> By the way, of both [wheat and barley] there are two different types concerning grains, spikes, shape and even the effects. In barley there is a 2-rowed type but also 3-, 4-, 5- and 6-rowed types. The Indian barley produces tillers. Some types show big and limp spikes, others are smaller and dense. Even the seeds of barley are either more round and smaller or longer and bigger. Some are white, others are reddish. (In A. F. HORT, Theophrastus, *Historia Plantarum*, de signis, de odoribus. London, 1916, p. 213.)

In *De Causis Plantarum,* THEOPHRASTOS discusses the problem of heritable modifications as follows:

> Sometimes changes happen on fruits sporadically, rarely on entire trees, and this the prophets call miracles, and it is seen as malformation against the nature, when it is happen in few cases but not if it occurs more frequent. Some herbs return to the wild shape too when they are without care; tiphia and zea [einkorn and spelt wheat] however turn into wheat within three years if they are sown without glumes, and this is associated with changes of growth conditions on site; because the annuals also change by time. (In B. EINARSON, *Theophrastus, De Causis Plantarum,* London, 1976, p.178.)

> Possibly the local conditions produce the plant species, either all or some of them. Because they unify the species or differentiate them with respect to usefulness as in the Thrakian wheat the

polyglumeness and late germination, both is caused by winter frosts. Therefore, in the other (regions, the author) the early sown Thrakian wheat comes up late and grow late and in turn of other (regions) sown germinates late. Then the habit has become like nature. (p. 217)

Herewith the problem of "transmutatio frumentorum" was named for the first time and with great emphasis. This problem caused many discussions throughout medieval times until the present (DITTRICH, 1959).

The term "inheritance" was already in use, though with a completely different meaning. In antiquity, it meant the production of descendants of the same type and with similar features. In this context, there were already "eugenic" measures for improving governmental staff.

In *Historia Plantarum,* there is a detailed description of plant flowers. It was recognized that some flowers were sterile. In cucumber vines and lemon trees, those flowers exhibiting an outgrowth as a distaff are sterile; those that lack such outgrowths are fertile. This seems to be the oldest note on a *carpel* (JESSEN, 1864).

W. CAPELLE (1949) investigates the botanical considerations of THEOPHRASTOS concerning the origin of plant malformations. THEOPHRASTOS pointed to three types of malformations:

1. Malformations caused by human treatment (i.e., agricultural and horticultural activities and/or missing care)
2. Malformation caused by environmental effects (e.g., temperature, wind, drought, etc.)
3. Malformations caused by the plant itself

According to his philosophy and the philosophy of antiquity—nothing happens without reason in nature—the latter type of malformations was classified as "incident against the nature."

The herbal *De Materia Medica* by P. DIOSCORIDES of Anazarba, Turkey (ca. 40-90 AD), a Greek physician serving in the Roman army, was written in the year 65 AD and is one of the most famous books ever written. It was referred to, copied, and commented on for 1,500 years. DIOSCORIDES did make an effort to systematize knowledge with plants originally grouped by form and origin, a practice continued until >>>LINNAEUS. In 1700, J. P. de TOURNEFORT (1656-1708) developed the first generic classification of plants by distinguishing clearly between genus and species, as published in his *Institutiones Rei Herbariae.*

Although the Romans historically did not do much to advance the understanding of biological inheritance, their performance with respect to applied inheritance, selection, breeding, and horticulture were outstanding. PLINIUS, the prolific author of *Historia Naturalis,* compiled a monumental encyclopedic treatment of science and ignorance. His coverage of the natural world is the best-known and most widely referenced source book on classical natural history. He records that KRATEUAS, a Greek herbalist and physician to MITHRIDATES VI, King of Pontus from 120 to 63 BC, described the nature of herbs and painted them in color, creating the first illustrated herbal.

On the other hand, the Romans demonstrated excellent knowledge of crop varieties. The Italian and native of Spain L. J. M. COLUMELLA (4 to ca. 70 AD) described, in his book *De re Rustica,* several wheat varieties, including *"siligo"* wheat *(Triticum monococcum).* Four varieties of spelt wheat *(T. spelta)* differentiated by color, quality, and grain weight were known. He also recommended the single-spike selection for cereals. PLINIUS characterized 15 beech, 15 olive, 4 pine, 4 quince, 7 peach, 12 plum, 30 apple, 41 pear, 29 fig, 18 chestnut, 11 hazelnut, and 91 grape varieties. COLUMELLA mentions that there are as many grape varieties as grains of sand in the Sahara desert. He knew that grapes grown in other than their native habitat lose their "character." M. T. VARRO (116-27 BC), P. V. M. VIRGIL (70-19 BC), and COLUMELLA published instructions for the best seed preparation. VARRO claimed:

> In order to have the best seeds for sowing, the best spikes have to be separately threshed. (In A. MAZZARINO, *Catonis de Agri Cultura,* Teubner Verl., Leipzig, 1922, p. 423.)

This could be viewed in context with a single-spike selection and positive *mass selection* of recent breeding methods. The rules given by VARRO are supplemented by COLUMELLA:

> I will add the instruction that immediately after harvest one has to care about good seed on the threshing floor. Namely, one

must, as CELSUS [25 BC-50 AD, a Roman encyclopedist] correctly mentioned, when cereals harvested of reasonable quality, the good spikes collect individually and thus look after the future. When better seeds were harvested, so grains have to winnowed in a vessel and the big and heavy once, which sink down to the bottom by winnowing, are kept for sowing. Of course, from a strong grain comes a good crop, a weak one brings a bad. (*De Arboribus;* in H. B. ASH, E. S. FORSTER, and E. HEFFNER, Loeb Classical Library, W. Heinemann Ltd., London, 1968, book III.)

VIRGIL acknowledged:

I have seen how cereals run wild even given greatest care, if not every year the best grains are selected by hand. (In H. R. FAIR-CLOUGH, *The Georgics.* Loeb Classical Library, W. Heinemann Ltd., London, 1974, book I.)

The best seed is the one-year-old one, less good the two-year-old, the three-year-old is the worst, and even older does not come up at all. For all crops the same rule is valid: what lies first on the threshing floor is kept for sowing because it is the best and the most heaviest. There is no other kind of differentiation. Spikes sat seeds only with gaps are drawn away. The best seed is reddish and has the came color if it is bite through with teeth, bad is that grain, which is white inside. (PLINIUS, in H. Rackham, *Natural History.* Loeb Classical Library, W. Heinemann Ltd., London, 1971, book XVIII.)

How precisely the Romans recorded and worked is proved in the quotation of COLUMELLA in *De Arboribus:*

In order to get good cuttings from good grape rootstocks, those rootstocks that show big, faultless, and ripen fruits are labeled with a mixture of vinegar and red chalk during harvest, which is not washed away by rain, and continues this procedure three or more years if the rootstock remains lasting good. Then one has sufficient evidence that the variety is excellent, and the quality and amount of berries is not incidentally influenced by a suit-

able vintage. (In H. B. ASH, E. S. FORSTER, and E. HEFF-
NER, Loeb Classical Library, W. Heinemann Ltd., London,
1968, book III.)

Those ancient approaches remind us of modern long-term and
multilocation testing—a testing of breeder's strains and varieties over
a more or less long period and on several geographically different
sites—in order to estimate the adaptive environmental response and/
or performance stability.

2.5. ARABIC AGRICULTURE

The Arabs were key in the exchange of crops between the East and
West. From India they brought sugarcane, rice, spinach, artichokes,
eggplants, orange, lemon, coconut, banana, and Old World cotton;
from Africa they introduced the watermelon and sorghum; and from
the Middle East the hard wheat *(Triticum durum)* and artichokes. The
flour of hard wheat is known as semolina, from the Arab "semoules"
(the German word *"Semmel"* meaning "roll" in English, derives
from the same Arab word). Having less gluten, it is not good for
bread but can be used for a number of processed products including
couscous and various pastas, such as spaghetti and ravioli.

The land of the Arab countries was dry. Thus, the great contribution of Arab agriculture was the introduction of summer irrigation,
which greatly intensified cultivation. The traditional Mediterranean
agriculture was based on winter or spring production. This partially
explains the incentive of Rome to conquer territory to import food-
stuffs. The Romans managed the shortage of water by fallow, in
which some land was not farmed so as to conserve moisture. When
the Arabs moved into Europe[4] they carried their irrigation technology
with them, allowing cultivation in the dry summers. The remains of
the irrigation technology can be seen today in southern Portugal,
where wells dot the countryside and primitive water-lifting devices
have only recently been electrified. Arabs introduced the Indian and
African summer crops and employed fertilization, including manure
as well as ground bones, crop residues, ashes, and limestone. They
introduced sugarcane and refined techniques for sugar manufacture.

2.6. MEDIEVAL AND RENAISSANCE AGRICULTURE IN EUROPE

The breakdown of the Roman Empire in the 6th century resulted in the destruction of the large cities, but it left rural areas and organizations relatively intact. There was a decline in knowledge in the Occident and a period of regression. During the next 600 years, the center of intellectual thought shifted to the Moslem world. Moslem and Jewish scholars who collected and translated the manuscripts from antiquity and developed scientific and technological schools of learning translated the manuscripts of antiquity from Greek to Arabic. In the West, vestiges of scholarship persisted in the monastery libraries, where learning was kept alive.

A feudal society developed, which involved the relation between land and the people who owned and worked in it. The lord of the manor, who owned the land, was served by the tenant farmers or vassals, who offered homage and fealty and owed a debt of labor. The cultivation regime was rigidly prescribed. The arable land was divided into three fields: one sown in the autumn with wheat or rye; a second sown in the spring with barley, oats, broad beans, or peas; and the third left as fallow. The fields were laid out in strips distributed over the three fields and without hedges or fences to separate one strip from another. About the 8th century, a four-year cycle of rotation of fallow appeared. The annual plowing routine on 100 ha would be 25 ha plowed in the autumn, 25 ha in the spring, and 50 ha of fallow plowed in June. These three periods of plowing, over the year, could produce two crops on 50 ha, depending on the weather.

Agriculture was the principal source of wealth in Medieval Europe. In the typical system, the land was divided by use into traditional categories: cultivated fields for grains, pulses, and fodder; meadows for grazing cattle, sheep, and pigs; and forests, which provided timber and game but were usually restricted to the lords. Annexed to the manors or tenant houses were kitchen gardens that provided fruit, vegetables, and herbs. Typically, the unfenced "open" fields were divided into strips allocated between villagers such that each had a portion of good and poor ground. Each worked his own strips and in addition provided labor to the lord. Systems of crop rotation were developed with crops one year and fallow every other or every third year. The system became entrenched, with little initiative left to individuals.

The Physica of HILDEGARD of Bingen (1099-1179) was the first book in which a woman discusses plants in relation to medical properties and the earliest book on natural history in Germany. She was a mystic, prolific author, and abbess of a Benedictine convent. *The Physica* was to have a great influence on the German botanists of the 16th century. In it the presence of beet *(Beta vulgaris)* was mentioned occurring at that time in Germany.

In many gardens inside monasteries, all sorts of plants were grown. In 1170, Cistercian monks from Burgundy mentioned the "Borsdorfer Apple," possibly the oldest variety. Cistercian monks distributed the pear tree during the 12th century from Burgundy to Bedfordshire (United Kingdom). They bred the first variety from a Burgundian rootstock. About 400 years later, in Italy, 232 varieties of pear were known. In 1254, the quince *(Cydonia oblonga)* was presumably introduced by the crusader ELEANOR of Castile and cultivated in the monks' gardens.

ALBERTUS MAGNUS (1193 or 1207-1280), whose German name was Albert von Bollstädt, was an early member of the Dominican Order of Preaching Friars, which established scholarly houses in the European centers of learning. He was responsible for the translations of ARISTOTELES and THEOPHRASTOS, and he influenced the revival of botanical and horticultural information based on the writing of antiquity in his work *De Vegetabilibus Libri* (1256). Most medieval writers thereafter drew inspiration from the Dioscoridian–Plinian tradition, although the description increasingly drew on the delineation of living plants and efforts were made to reconcile ancient writings with native flora. Issues such as inheritance and modificability are discussed in the fifth book of *De Vegetabilibus Libri*. The unification of one plant species with another and the conversion of one species into another are particularly treated. It appears again as one of THEOPHRASTOS's assumptions that, for example, rye can be turned into wheat after the second or third year after sowing, as wheat can convert into spelt wheat and vice versa again into rye. This theory was believed until modern times. A. MAGNUS saw the cause of conversion in the metabolism of soil, where plants take up the nutrients—that is, that potency of matters of one organism may be transferred to another. He believed that mushrooms could develop from the stubs of oaks or birches and that grasses can develop from rotten substances. Another type of conversion is realized by grafting of plants (>>> LYSENKO; see Chapter 3.9). For example, if plums are grafted onto willow, seedless fruits will originate. Grapevine scions grafted on cherry, apple, or pear rootstocks result in scions showing the same ripening time as the rootstock. The conversion of the wild into the crop plants was explained as the influence of nutrition needs, soil preparation, environment, sowing time, and so forth. MAGNUS did not recognize heritable changes, although he accepted malformations in animals that are inherited.

By the 14th century, a trend toward more naturalistic drawings became apparent, suggesting the influence of the Renaissance in art and ideas. Plants were no longer slavishly copied from past manuscripts but instead were redrawn afresh from the local flora. In the late Middle Ages, the concept of "doctrine of signatures" developed in which the inner qualities of plants were thought to be revealed by external signs. This was codified by mystical writers, such as the Swiss physician PARACELSUS (1493-1541) and G. della PORTA (1539-1615), the author of *Phytognomonica Octo Libris Contenta* (Napoli, 1588). Thus, long-lived plants would lengthen a man's life while short-lived plants would abbreviate it. Yellow sap would cure jaundice; plants with rough surfaces would cure diseases that destroy the smoothness of skin. Plants that resembled butterflies would cure insect bites. The concept was that medical plants were stamped with a clear indication of their uses. A 17th-century dispensary explains the concept:

> The powers of hypericum [St. John's wort] are deduced as follows: I have oft declared how by the outward shapes and qualities of things we know their inward virtues which God hath put in them for the good of man. So in John's wort we take notice of the form of the leaves and flowers, the porosity of the leaves, the veins. 1. The porosity roles in the leaves signify to us, that his herb helps both inward and outward holes or cuts in the skin. 2. The flowers of Saint John's wort, when they are putrefied they are like blood; which teaches us, that this herb is food for wounds, to close them and fill the up. (In PARACELSUS, *Aphorismorum Aliquot Hippocratis Genuinus Sensus & Vera Interpretation*, M. Franck für G. Willer, Augsburg, 1568.)

This herb plant, St. John's wort, has more recently been subjected to breeding, since the hypericin is extensively used as a mood elevator and nerve calmer. In Germany, its 2005 sales were comparable to those of conventional antidepressants, which is largely due to the extensive clinical research and data on this plant.

The long period of medieval agriculture in Europe led to the present divisions of agriculture into agronomy, horticulture, and forestry. Agronomy became involved with open fields and meadows for the production of grain and fodder. The kitchen gardens of tree fruits and vine, vegetables, ornamentals, and herbs became the domain of horticulture. The forests for timber and game became a special purview of forestry. Initially the differences between agronomy and horticulture were based on crops and intensity of production. Of course, this system cannot be maintained in the tropics, where it is never clear where agronomy ends and horticulture begins.

With the age of printing, there was a tremendous demand for books of agricultural works, including farm management and vine culture. Consequently, agricultural technology of the period is available from the printed record. A particularly good example is *L'agriculture et Maison Rustique* of C. ESTIENNE and J. LIÉBAULT (Paris, France, 1564), which went through many printings starting in the late 1500s. It was translated into English by R. SURFET and published as the *Country Farm* in 1600 (London), enlarged in 1616, and went through a series of editions. These books are a tremendous source of information about crop improvement. The literature of the Middle Ages and Renaissance, which developed in the 14th century, has made it possible to develop a complete history of horticulture and agronomy. It was a difficult time in Europe because of climate change— the Little Ice Age—that reduced yields, the rise of diseases, and the increase in populations. There was a demand for new land. The discovery of America by C. COLUMBUS (1451-1506) was a part of those strategies. COLUMBUS discovered America in 1492 and described red pepper and allspice in his journal:

> We ran along the coast of the island, westward from the islet and found its length to be 12 leagues as far as a cape which I named Cabo Hermoso, at the western end. The island is beautiful . . . I believe that there are many herbs and many trees that are worth

much in Europe for dyes and for medicines; but I do not know them and this causes me great sorrow. There are trees of a thousand sorts, and all have their several fruits; and I feel the most unhappy man in the world not to know them, but I am well assured they are valuable. (In P. HALSALL, 2000, *Medieval Sourcebook: Christopher Columbus: Extracts from Journal of Columbus in His Voyage of 1492*.)

Two of his soldiers reported about people on the island of Cuba eating flour from seeds of new sort of grass (probably maize!). By this time, the American Indians had developed about 200 different varieties of flint maize. Dent maize appeared in northeastern Mexico. They have the advantage of a substantial amount of soft starch in their kernels surrounded by a thin outer layer of hard starch. Those maize were introduced to Spain in 1520. In 1596, the monk Romano PANE, one of COLUMBUS' crew, described tobacco for the first time, and how the Indians smoked it. Since 1550, the tobacco plant has been grown in Spain.

In 1539, the German H. BOCK (1498-1554) published the *New Kreutterbuch von Underscheidt, Würckung und Namen der Kreutter, so in Teuschen landen Wachsen* (Book of Herbs) in Strasbourg. It contained many descriptions of horticultural and agricultural plants, such as broad bean, pea, white lupin, and clover.

By the 16th century, human population was increasing in Europe and agricultural production was again expanding. The nature of agriculture and horticulture there and in other areas was to change considerably in the succeeding centuries. Several reasons can be identified. Europe was cut off from Asia and the Middle East by an extension of Ottoman power. New economic theories were being put into practice, directly affecting agriculture. In addition, continued wars between England and France, within each of these countries, and in Germany consumed capital and human resources. Colonial agriculture was carried out not only to feed the colonists but also to produce cash crops and to supply food for the home country. This meant cultivation of crops (e.g., sugar, cotton, tobacco, and tea) and stimulated detailed studies of plant development.

Concerning scientific progress of crop sciences, toward the end of 16th century we receive the first comprehensive description of a spontaneously occuring heritable change on a higher plant. The drug-gist and professor of botany at the University of Heidelberg (Germany), P. S. SPRENGER, found in his experimental garden in 1590 a new type of celandine *(Chelidonium majus)* with deeply lopped leaves. He called it *Chelidonium folio lanciniato*. The mutant form showed long-lasting stable inheritance without any segregation, as was clearly shown by >>> MILLER's (1768) later study, which spanned forty years. The spontaneous mutation in *Linaria* with formation of peloria flowers, which caused LINNÈ to doubt the unique creation of species, was not the only 18th-century example known from higher plants. Similar reports are given by MARCHANT (1719) in *Mercurialis annua* with laciniate habit and in *Fragaria vesca* with a simple oval leaf (DUCHNESNE, 1766).

2.7. PLANT BREEDING BY EXPERIENCE DURING THE 17TH THROUGH 19TH CENTURIES

The scientific revolution resulting from the Renaissance and the Age of Enlightenment in Europe encouraged experimentation in agriculture as well as in other fields. Cereal yields were still low, around 0.8-1.0 t/ha; higher were the yields of cabbage (about 25 t/ha). Therefore, fermented cabbage became a more popular diet in central Europe than cereal products. Trial-and-error efforts in plant breeding produced improved crops.

After the first step of unconscious influencing and improvement of cropped plants led to a conscious selection of the best, biggest, or most beautiful, the way was free for breeding by experience or plant breeding as an "art." Only few people practiced it using their individual knowledge and experience. Particularly in breeding ornamental plants, a breath of artistic interest determined the actions and success (such as tulip breeding and production in Holland). This form of breeding was devoid of scientific basis, but it developed the basic types of modern crop plants over more or less long periods of selection.

An impressive example is the wide variety of cabbages derived from a simple *Brassica* genotype. The gradual development can be followed by presentations in the old herb books. A second (similar) example can be given. In the daily market places of Mexico, a wide range of at least 25 different varieties of common bean *(Phaseolus* vulgaris) is offered: white beans for meat meals, black beans are eaten with tortillas, some only roasted or boiled, red beans suitable for soups, and so on. Over the centuries, a complex culture of beans developed. On the one hand it shows the habit of humans to select for color, taste, constitution, nutritional value, and storage ability; on the other, it demonstrates how a high degree of genetic variability within a crop species was permanently maintained. This situation dramatically changed with modern plant breeding of the 20th century, where uniform and standardized plant products are required by food technology and international trade. Therefore, it became a major task of national gene banks of the 21st century to maintain the genetic diversity of crop plants by collection, storage, and evaluation (see Chapter 4).

The foundation of botanic gardens during the 16th and 17th centuries did much in the way of advancing botany, horticulture, and agriculture (Table 2.1). At first these gardens were chiefly devoted to the cultivation of medical plants. This was especially the case at universities where medical schools existed. The first botanic garden was established at Pisa in 1543.

TABLE 2.1. Foundation of first botanical gardens in Europe.

Year	Location	Country
1543	Pisa	Italy
1543/1544	Padua	
1545	Florence	
1567	Bologna	
1590	Leiden	Holland
1579	Leipzig	Germany
1597	Heidelberg	
1625	Altdorf	
1605	Giessen	
1629	Jena	
1593	Montpellier	France
1620	Strasbourg	
1626	Paris	
1600	Copenhagen	Denmark
1655	Uppsala	Sweden
1621	Oxford	England
1730	Kew	

Although there is little information about plant breeding from this epoch, the results demonstrate the progress of breeding—even if the steps were small because workers did not recognize the difference between genetic and environmental variation. More detailed references are available only from the last centuries and only from countries where economic development matched up well with agricultural requirements. Thus it is not surprising that Holland, England, and Italy played dominant roles during this period.

2.8. INTEREST

In the 1670s, the Dutch >>> A. van LEEUWENHOEK (1632-1723) built the first microscopes with 250-fold magnification for plant and other studies; in 1665, >>> R. HOOKE (1635-1703, England) described the cell as the basic functional unit of organs; and in 1675, >>> M. MALPIGHI conducted anatomical investigations in plants and repeated the observations of HOOKE. In 1676, T. MILLINGTON (England) saw anthers function as male organs, and in 1670, >>> GREW (1641-1712, England) suggested the function of pollen and ovules. In 1823, J. B. AMICI (1784-1860) noted pollen tubes in flower studies. C. F. WOLFF (1733-1794), a German biologist, concluded from his studies on animals and plants that nutrition and growth of plants are caused by an "essential force" ("vis essentialis") that converts a homogeneous, clear, and glassy matter into new organs not previously present (with respect to an egg). His knowledge can be summarized as follows:

1. All living beings show cell structures as a common trait.
2. Plant and animal cells are basically similar.
3. The morphological appearances (fibers, vessels, etc.) are modifications of cells.

The notions of *constancy* and *variability* were the main biological phenomenons that were on his mind. WOLFF recognized two types of variability. The first is influenced by light, temperature, air, moisture, and nutrition. This variability, caused by the environment, he called "variation." He knew that plants brought from St. Petersburg (Russia) to Siberia subsequently change their habit—and when returned to St. Petersburg they took up their former shape. He obviously knew the idea of modification and its inheritability when he held a chairmanship of anatomy at the St. Petersburg Academy of Sciences after 1767:

> The variations are not inherited, and the outer shape itself and its internal structure would be constant if they would not be changed by random, not inheritable predicates. (In *Von der Eigenthümlichen und Wesentlichen Kraft der Vegetablischen Sowohl als Auch der Animalischen Substanz*, St. Petersburg, 1789, p. 41.)

In 1694, R. J. CAMERARIUS (1665-1721, Germany) published *De Sexu Plantarum Epistola*, which summarized all data on sexual knowledge and was the first demonstration of sex in plants. When dealing with mulberry and maize, he also suggested crossing as a method to get new types. Systematic plant breeding was established at this point. Similar findings by the American C. MATHER (1663-1728), who observed natural crossing in maize, date from 1716. He noted *xenia*—that is, where ears of yellow maize are planted next to red and blue maize, the yellow maize showed red and blue kernels during the first year. By this method, MATHER was able to describe spontaneous outcrossing in neighboring rows. Supporting results were later achieved in controlled experiments by P. DUDLEY (1724, see SWINGLE and WEBBER, 1897) and J. LOGAN (1739). In his book *Religio Philosophica*, MATHER also reported about spontaneous hybrids between *Cucurbita pepo* var. *ovifera* and *C. pepo* var. *condensa* or *C. pepo* var. *maxima*. He noticed bitter fruits when *C. pepo* var. *ovifera* was pollinated with *C. pepo* var. *condensa*.

But T. FAIRCHILD (1667-1729)—a versatile and talented gardener known for the introduction of exotic plants to England—created in 1717 the first proved artificial hybrid of carnation and sweet william *(Dianthus caryophyllus × Dianthus barbatus)*, commonly known as "Fairchild's mule." The hybrid was sterile and was vegetatively propagated for more than 100 years. E. LEHMANN (1916) mentioned a forerunner of >>> J. G. KÖLREUTER by reference of a Dutch botanist, E. MORREN—the Parisian postman N. GUYOT—who made plant hybrids in *Ranunculus* and *Primula* based on his knowledge of sexual propagation and the function of pollen grains. However, KÖLREUTER was the first to perform systematic studies on plant hybrids. In 1761 he used tobacco *(Nicotiana rustica × N. paniculata,* and other hybrids of *Dianthus, Matthiola, Hyoscyamus, Verbascum, Hibiscus, Datura, Cucurbita, Aquilegia,* and *Cheiranthus)* to demonstrate that hybrid offspring received traits from both parents (pollen and ovule transmit genetic information) and were intermediate in most traits. In a paper of 1763, he concluded that mother and father equally and specifically contribute to the character of hybrid:

> For production of each plant in nature two equally liquid matters of different kind are needed for unification, which are predetermined by the creator of all things. (In Vorläufige Nachricht von Einigen das Geschlecht der Pflanzen Betreffenden Versuchen und Beobachtungen, Gleditschischen Handlung, Leipzig, p. 37.)

He demonstrated the identity of reciprocal crosses and mentioned hybrid vigor (now called *"heterosis;"* see Chapter 3.4) and segregation of offspring (parental and nonparental types) from a hybrid.

Nevertheless, the hybridization experiments and the "change of species" by backcrossing were not recognized and were even denied by KÖLREUTER's contemporaries. Thus, as in other sciences, the history of genetics and plant breeding demonstrate that important discoveries are often not recognized or long ignored. The rejection of the works of CAMERARIUS, KÖLREUTER, and SPRENGEL was a big failure of an epoch in which the dogma of invariability of species was still not overcome. SPRENGEL also published in 1793 a description of the role of insects in pollination of angiosperms and discussed the flower structure and color in relation to pollination and insect habits. These observations were prerequisites for the development of plant genetics.

On the other hand—and unimpressed of the academic discussions—practical breeders such as the American C. M. HOVEY, the "Father of the American Strawberry," developed the first fruit cultivar of strawberry using controlled pollination in 1834.

Only by end of the 18th century did progressive English, French, and German plant breeders begin systematic hybridization experiments in order to increase the crop yield and make experimental gardens for their observations. In addition to production of plants for consumption and trade, they carried out numerous crosses with the aim of new mixtures of traits. In 1727, under the supervision of >>>

P.-L.-F. VILMORIN, a French company practiced the *pedigree method* for breeding sugar beet. The Vilmorin-Andrieux et Cie Seed Company contributed extensively to the development of plant breeding knowledge and improved cultivars for over 260 years. In 1785, J. B. van MONS started systematic selection of horticultural plants. In his seed catalog of 1825, he listed 1,050 cultivars (DECAISNE, 1855). In 1825, the American farmer J. LORAIN recognized the possibility of maize hybrid seed, and in 1830, "Red May"—the first bread wheat cultivar selected in the United States—was released.

In 1779, >>> T. A. KNIGHT (1759-1838), president of the Horticultural Society of London from 1811 to 1838, emphasized the practical aspects of hybrids and worked on improving frost resistance and yields in different plants (grape, apple, pear, plum, garden pea, and probably even wheat) rather than on inheritance. However, he noted the advantage of outcrossing to produce new forms of crops and that male and female parents contribute equally to the resulting hybrid, with segregation following in the next generation. He made also the first wheat variety crosses in Europe. His pea crosses from 1799 to 1823 are most important. He realized—as did >>> G. MENDEL much later—several varieties with distinct character via self-fertility, flower isolation, and simple cultivation. He carefully emasculated the flowers and used them as control flowers without pollination. A luxuriance of hybrids was observed early. As first researcher, KNIGHT found the dominance of the grey seed color over white. After backcrosses of white-seeded plants to the hybrid, he obtained segregated white-seeded and grey-seeded offspring—though, of course, without calculating the numerical proportion.

Similar data were achieved by the Italian pomologist >>> G. GALLESIO using carnations. In his book *Teoria della Riproduzione Vegetale* in 1816 (Pisa, Italy), he explained his observations about crossing experiments:

> I fertilized white flowering carnations with pollen of red flowering plants and vice versa; the carnations I cultivated from the seeds showed flowers with mixed color. (p. 24)

> Therefore with their unification, which is not natural, arises an irregularity of their effects, and those soonly show the character

of the one origin or the other, according to circumstances the one dominates. (p. 30)

MARTINI (1961) remarked that GALLESIO was the first to use the term "dominate," although ROBERTS (1929) noted in his book *Plant Hybridisation Before Mendel* that >>> M. SAGERET, a researcher and agronomist and member of the Société Royale et Centrale d'Agriculture de Paris, used it for the "first" time in 1826 (i.e., about ten years later!).

Inspired by KÖLREUTER, >>> W. HERBERT, an English vicar of Manchester and contemporary of T. A. KNIGHT, suggested that the character of winter hardiness is inherited. HERBERT investigated the fertility of hybrids (whether fertile or sterile) as a measure of the relationship between species and varieties. Although he recognized the importance of hybridization for horticulture and agriculture—as did KNIGHT for improvement of fruit trees—he ignored crossability as a measure for phylogenetic relationships:

> The only thing certain is, that we are ignorant of the origin of races; that God has revealed nothing to us on the subject; thereon; but we cannot obtain negative proof, that is, proof that two creatures or vegetables of the same family id not descend from one source. But can we prove the affirmative; and that is the use of hybridizing experiments, which I have invariably suggested; for if I can produce a fertile offspring between two plants that botanists have reckoned fundamentally distinct, I consider that I have shown them to be one kind, and indeed I am inclined to think that, if a well-formed and healthy offspring proceeds at all from their union, it would be rash, to hold them of distinct origin. (In On hybridization amongst vegetables, *J. Hort. Soc.,* London 2, p. 67.)

Later, however, G. GODRON in his 1844-1863 experiments used the character of sterility of interspecific hybrids as evidence that the parents were distinct species. But the German botanist A. F. WIEGEMANN (1771-1853), a member of the "Kaiserliche Akademie der Naturforscher" (Leopoldina), had already refuted several prejudices of his contemporaries in a paper titled "Über die Bastarderzeugung im Pflanzenreiche" (1828). His thesis can be summarized as follows.

1. There is hybridization in the plant kingdom. It gives evidence for sexuality in plants. The influence of the male parent by transfer of the pollen to the stigma can be seen on several traits.
2. The pollen substance is taken up by the secretion of the stigma and is transferred through pistil to the carpel by mediation of the cell tissue.
3. Some hybrids bear a slight resemblance to the father; some are intermediate. More frequently, the hybrids take after the mother.
4. LINNÉ's hypothesis is disproved that sexual organs of hybrids are from the mother and foliage and habit from the father.
5. Hybrids are fertile if they derive from crosses between varieties and species.
6. Individuals deriving from hybrids often strongly deviate from the characters of the mother or father and show heterogeneous habits.
7. The closer the parental plants, the easier is the production of hybrids. The best is the crossability between varieties, followed by species of one genus and plants of different genera.

Based on this, studies the Dutch Academy of Sciences at Haarlem in 1830 initiated a competition: "What teaches the experience concerning creation of new species and varieties by artificial fertilization of flowers by pollen, and which crop and ornamental plants can be produced and multiplied in this way?" The winner, C. F. von GÄRT-NER (1772-1850), received the award belatedly in 1837. Within the period 1827-1849, he made the most extensive crossing experiments to date, with 10,000 crosses in 700 species from 80 genera. By analysis of more than 9,000 experiments over a period of seven years, he briefly reported to the Academy. Although his statements are similar to WIEGEMANN's, he concluded that the origin of new varieties after hybridization simply confirms the irrefutable border of species that cannot be overcome. As deserving as his work was, it brought him only criticism seventeen years later. MENDEL (1865) wrote in the introduction to his famous paper "Versuche über Pflanzen-Hybriden" and in a letter to C. von NÄGELI that GÄRTNER's experiments were difficult to explain since clear descriptions of trials and diagnosis of hybrids were missing and since the characterization of hybrid traits was perfunctory. C. NAUDIN (1863), in his paper

"Noevelles recherches sur l'hybridité dans les végétaux," described the separation of essences in various proportions in germ cells and recombination in the offspring of a cross when he used *Datura* hybrids for his rather complex experiments. He thought that the whole plant was a simple mosaic of the two parents.

Since the beginning of the 19th century, experiments with pea were *en vogue*. As did the English gardener T. A. KNIGHT in 1789, >>> J. GOSS chose varieties of pea as subject for hybridization experiments. He reported in 1822 to the Secretary of the Horticultural Society of London about his crosses of "Blue Prussian" with "Dwarf Spanish" peas:

> . . . that these white seeds had produced some pods with all blue, some with all white, and many with both blue and white peas in the same pod. (In On the variation in the colour of peas, occasioned by cross-impregnation, *Transact. Hort. Soc.*, London 5, p. 234.)

He separately grew blue and white seeds. Then blue seeds produced only plants with blue seeds; however, the white seeds gave plants with white and blue seeds in the pods. In 1824, A. SETON mentioned a cross of the green-seeded pea "Dwarf Imperial" with a white-seeds variety:

> . . . they were all completely either of one color or if the other, none of them having an intermediate tint. . . . (In On the variation in the colour of peas from cross-impregnation, *Transact. Hort. Soc.*, London 5, p. 237.)

Both GOSS and SETON thus observed in their experiments the phenomenon of *dominance* and segregation, as had T. E. KNIGHT. In 1866 and 1872, a few decades after SETON's pea experiments, the breeder T. LAXTON again carried out hybridizations with pea that confirmed the findings of KNIGHT and GOSS: dominance of one of the parental traits (e.g., seed color or seed shape) and segregation of the traits in the second generation. All segregants were separately harvested and checked in further generations. The result reflected the grouping of segregants with two associated features in the sense of MENDEL. As LAXTON (1872) explained:

I have noticed that a cross between a round white and a blue wrinkled pea, will in the third and fourth generations (second and third year produce) at times bring forth blue round, blue wrinkled, white round, and white wrinkled peas in the same pod, that the white round seeds when again sown, will produce only white round seeds, that the white wrinkled seeds will, up to the fourth or fifth generation, produce both blue and white wrinkled and round peas, that the blue round peas will produce blue wrinkled and round peas, but that the blue wrinkled peas will bear only blue wrinkled seeds. (In Notes on some changes and variations in the offspring of cross-fertilized peas, *J. Roy. Hort. Soc.*, London 3, p. 13.)

As a breeder, LAXTON also knew that maximum variability could be expected in the third and fourth generation after crossing and that it was impossible to obtain stable new varieties before the third and fourth generation (although a setback to parental traits could happen earlier). The French breeder >>> P.-L.-F. de VILMORIN, the one who introduced the principle of individual *testing of progeny* in 1860, stressed the value of self-pollination in breeding stable cultivars of wheat and sugar beet and also confirmed LAXTON's data on lupin *(Lupinus hirsutus)* in experiments between 1856 and 1860. His son, H. de VILMORIN (1843-1899), was especially interested in new combinations *(= recombination)* of (formerly) associated characters; in 1878, he did the first extensive work with interspecific crosses in cereals *(Triticum vulgare × T. polonicum* and *T. durum).*

Many breeders of the 19th century were aware of spontaneous variations and used them for the selection of new strains. >>> K. RÜMKER (1889) mentioned some examples in his book *Anleitung zur Getreidezüchtung auf Wissenschaftlicher und Praktischer Grundlage* (Instruction for Cereal Breeding on Scientific and Practical Basis). Obviously, the Frenchman J. le COUTEUR systematically selected deviating segregants from hybrid combinations (see the following text). In 1841, R. HOPE described a "Fenton" wheat derived from a single plant selection. In 1873, the Scottish breeder >>> P. SHIRREFF gave many examples. He developed more than eleven varieties of oat and wheat based on the selection of spontaneous variants:

"Mungowells wheat," based on one plant from 1819 and stable after the fourth generation
"Hopetoun oats," based on one plant from 1824
"Hopetoun wheat," based on one plant from 1832
"Shirreff oats," origin unknown
"Shirreff's bearded red wheat" (by 1860 distributed for sale)
"Shirreff's bearded white wheat" (by 1860 distributed for sale)
"Pringles wheat" (by 1860 distributed for sale)
"Early fellow oats" (by 1865 distributed for sale)
"Fine fellow oats" (by 1865 distributed for sale)
"Long fellow oats" (by 1865 distributed for sale)
"Early Angus oats" (by 1865 distributed for sale)

SHIRREFF first established nursery plots with seventy Scottish families by 1857. He suggested crossing parents with desirable charac teristics to obtain progeny of value. By 1819 he had observed in his wheat field a particularly vigorous type. He fertilized it, removed all surrounding plants, and harvested 63 spikes with 2,473 grains. After propagation over many years, the offspring showed specific characteristics that deviated from the original population; nowadays it would be called "mutant." In this way, SHIRREFF developed several other wheat varieties. In rye, similar approaches were reported by MARTINI (1871), who tried to fix the three-flowered spikelets as BLOMEYER had with the variety "Leipziger Roggen." WOLLNY (1885) was successful with breeding the "Igelroggen" (hedgehog rye) and the "Laxblättriger Roggen" (lax-leaf rye) based on single plant selection. In 1883, >>> RIMPAU (1899) found—in two-rowed barley and rye plants with branched spikes—a white-glumed, an awned, and a compactum plant in a red-glumed awnless landrace of wheat. Of particular breeding value was the awnless "Anderbecker Hafer" (Anderbecker oats) of BESELER (Germany), originating from a single plant selection within the awned "Probstei Hafer." DRECHSLER found on the experimental field of Göttingen (Germany) an awned type of the squarehead wheat, and METZGER (1841) reported twelve variants of maize.

With the beginning of industrialization in Europe and America, examples for targeted improvement of crop plants and scientific inventions are numerous: The increased demand for sugar led to the breeding of sugar beet as an industrial crop. After the 1747 discovery by the German chemist A. S. MARGGRAF (1709-1782) that beets contain sugar, >>> F. C. ACHARD (1753-1821), also a German chemist, was the first person that selected beet plants successfully for sugar production. Between 1786 and 1800, he grew about 22 varieties of beet on an experimental field at Kaulsdorf near Berlin for testing the sugar content. He is considered the founder of the sugar beet industry. By crossing of the "Weisse Mangoldrübe" (Zuckerwurzel = sugar root) and the "Roter Mangold" (Rübenmangold or Futterrübe = fodder beet), he was able to select for genotypes showing high sugar content. It was the birth of a new crop plant. The first name of sugar beet was called "Schlesische Rübe." The first cropping of sugar beet started in central Germany around Halberstadt and Kleinwanzleben, where heavy fertile soils were available (it remained the center of sugar beet production of Germany). Later, ACHARD founded the first sugar beet factory of the world at Cunnern (Silesian, Germany/Poland).

Fodder beets were already bred by W. von BORRIES in Eckendorf (Germany) since the beginning of the 19th century. In 1849, he grew twenty varieties in a nursery for morphological and yield comparison. In 1840, W. RIMPAU, Schlanstedt, and F. KNAUER, Gröbers (Germany) introduced the method of mother beet selection and its pedigree testing in order to speed up the breeding procedure for yield performance and sugar content of the "Magdeburger Zuckerrübe." The method became even more efficient when, in 1862, M. C. RABBETHGE (1804-1902) and J. GIESEKE (1833-1881), the founders of the "KWS" International Seed Company, applied light polarization (discovered in France by F.-B. BIOT, 1811) of beet sap for screening and determining sugar content.

Other important inventions during this period can be listed as follows. In 1853, the American E. W. BULL produced "Concord" grapes by hybridizing European cultivars and wild grapes of New England. In 1855, R. L. VIRCHOW (1821-1902, Germany) established that the egg cell of one generation comes from the egg of the previous generation and demonstrated the "continuity of heredity." In 1858, W. HOFMEISTER (1824-1877, Germany) discovered the female gametophyte and the law of change of generations. H. S. BIDWELL (1867, United States) first introduced detasseling of maize in his breeding program as a measure to control pollination. In 1875, the first hybrid oat ("Pringles Progress") was released in the United States. In the years 1875-1877, STRASBURGER (Germany) gave the first adequate description and drawing of chromosomes and showed that nuclei arise only from nuclei; he suggested the terms *"gamete"* and *"chromosome."* In 1878-1881, >>> W. J. BEAL (United States) proposed crossing maize cultivars in order to increase the yield of commercial types. In 1879-1880, F. HORSFORD (United States) selected the first known hybrid cultivar of barley; in 1885-1887, >>> A. WEISMANN (Germany) wrote twelve essays on heredity and evolution that became important in directing the trend of biology and thought. He summarized current knowledge and pointed out the broad implications of "continuity of germplasm," "equal inheritance from parents," and "nonheritability of acquired characters." In 1884, E. STRASBURGER (Germany) demonstrated fertilization and showed the fusion of two nuclei to form the zygote; in 1888, he demonstrated reduction division in plants. In 1886, S. M. SCHINDEL developed the first hybrid winter wheat "Fulcaster" in Maryland (United States) from a cross *Triticum turgidum* ssp. *durum,* cv. Fultz and *T. aestivum* ssp. *aestivum,* cv. Lancaster land race. In 1890, the breeder W. J. FARRER (Australia) ran an extensive wheat-breeding program using hybridization for selection of rust resistance. In the same year, W. M. HAYS (United States) used *centgener test* and pedigree selection on wheat and oats after starting plant breeding in Minnesota in 1888. From the beginning, he used the individual-plant method of selection. "Improved Fife," "Minnesota 163," "Minnesota 169," and "Haynes Bluestem" were valuable new varieties of spring wheat selected by this method. In 1891, W. A. KELLERMAN and W. T. SWINGLE did the first counts of segregation for the starchy gene on maize ears. REID's "Yellow Dent Maize" won the grand prize as "the world's most beautiful corn" at the World's Colombian Exposition in 1893 at Chicago; REID's maize became a major force in Midwestern agriculture and an important parent to modern hybrid maizes. In 1899, the American C. G. HOPKINS (the mentor of >>> EAST when he came as chemist to study the oil and protein content of maize) described the ear-to-row selection method *(half-sib progeny testing).* However, major advances in plant breeding followed the revelation of MENDEL's discovery. Breeders brought their new understanding of genetics to the traditional techniques of self-pollinating and cross-pollinating plants.

Chapter 3

MENDEL's Contribution to Inheritance and Breeding

The prevailing view of heredity during the middle of the 19th century assumed two gross misconceptions: the acceptance of a blending of hereditary factors and the heritability of acquired characters. Evidence for a particulate basis for inheritance, such as the reappearance of ancestral traits, was common knowledge but considered exceptional. In fact, all the "discoveries" attributed to MENDEL, such as the equivalence of reciprocal crosses, dominance, uniformity of hybrids, and segregation in the generation following hybridization, are gleanable from the pre-Mendelian literature.

The many achievements in sciences and agriculture by the middle of the 19th century, the numerous hybridization experiments (very often with pea varieties) and >>> C. DARWIN's paper on "Origin of Species" demonstrating genetic variation, inbreeding, sterility, and differences in reciprocal crosses can be seen as the fertile lap from which G. MENDEL, an Austrian Augustinian monk in the monastery of Brünn (now Brno, Czech Republic), received some of his inspiration.

DARWIN's explanation of evolution by natural selection became a well-established theory in the years following publication in 1859 despite any factual evidence to explain either the nature or the transmission of hereditary variation. DARWIN, aware that blending inheritance led to the disappearance of variation, relied on the inheritance of acquired characters to generate the variability essential to his theory. His clear but inaccurate formulation of a model of inheritance,

Concise Encyclopedia of Crop Improvement
© 2007 by The Haworth Press, Inc. All rights reserved.
doi:10.1300/5891_03

which involved particles ("gemmules") passing from somatic cells to reproductive cells, exposed the lack of any factual basis. The concept was merely a restatement of views dating from HIPPOCRATES in 400 BC and endlessly reformulated.

In his paper "Versuche über Pflanzen-Hybriden," G. MENDEL found that the plant's respective offspring retained the essential traits of the parents and therefore were not influenced by the environment. This simple test gave birth to the idea of heredity. The proof of the particulate nature of the elements was made possible by the nature of the plants and traits studied. The traits chosen were contrasting (e.g., yellow versus green cotyledon, and tall versus dwarf plant) and constant (i.e., true breeding) after normal self-pollination in the original lines. In seven of the eight characters chosen for study, the hybrid trait resembled one of the parents; in one character, bloom date, the hybrid trait was intermediate. MENDEL called traits that pass into hybrid association "entirely or almost entirely unchanged" dominating, and the latent traits were termed recessive (recidivous) because the trait reappeared in subsequent crosses. He saw that the traits were inherited in certain numerical ratios. Thus, he then came up with the idea of dominance and segregation of genes and set out to test it in peas. The phenomenon of dominance was clearly not an essential part of the particulate nature of the genetic elements but was important to classify the progeny of hybrids. The disappearance of recessive traits in hybrids and their reappearance, unchanged, in subsequent generations was striking proof that the elements responsible for the traits (now called genes) were unaffected in their transmission through generations.

When the hybrids of plants differing by a single trait were self-pollinated, three-fourths of the progeny displayed the dominating trait and one-fourth the recessive. In the next self-generation, progeny of plants displaying the recessive trait remained constant (nonsegregating), but those with the dominant trait produced one of two patterns of inheritance. One-third was constant, as in the original parent with the dominating trait, and two-thirds were segregating as in the hybrid. Thus, the 3:1 ratio in the first segregating generation (now called the F_2 generation) was broken down into a ratio of 1 (true breeding dominant) : 2 (segregating dominant as in the hybrid) : 1 (true breeding recessive). The explanation proposed was that theme parental plants had paired elements (e.g., *AA* or *aa*, respectively) and the hybrids of such a cross were of the constitution *Aa*.

It took seven years to cross and score the plants to the thousand to prove the laws of inheritance. From his studies, MENDEL derived certain basic laws of heredity:

1. Hereditary factors do not combine but are passed intact.
2. Each member of the parental generation transmits only half of its hereditary factors to each offspring (with certain factors "dominant" over others).
3. Different offspring of the same parents receive different sets of hereditary factors.

MENDEL, in his paper, spoke about the "law of combination of different characters" and talked about "the law of independent assortment." He implied that the segregation of factors occurred in the production of sex cells. Thus, he conceptualized a genetic theory that created a new view on biology. The standard approach to unravel the mysteries of heredity was to analyze complex characters usually from wide crosses, a method that had failed for 2,000 years and was to continue to fail even when applied by the combined talents of >>> F. GALTON and K. PEARSON (see Chapter 3.10).

MENDEL succeeded because of his approach. His goal was grand, being no less than to obtain a "generally predictable law" of heredity. His previous crosses with ornamentals had indicated predictable patterns, and his assumption was that laws of heredity must be universal. He had reviewed the literature and noted:

> . . . that of the numerous experiments, no one had been carried out to an extent or in a manner that would make it possible to determine the number of different forms in which hybrid progeny appear, permit classification of these forms in each generation with certainty, and ascertain their numerical relationships (in Versuche über Pflanzen-Hybriden, *Verhandlungen des Natur-forschenden Vereines*, Abhandlungen, Brünn 4, p. 34, 1866).

The experimental organism, the garden pea, was once again a perfect choice. MENDEL procured thirty-four cultivars from seedsmen, tested their uniformity over two years, and selected twenty-two for hybridization experiments. Results from preliminary crosses indicated that common traits were transmitted unchanged to progeny but that contrasting traits may form a new hybrid trait that changes in subsequent generations. A series of experiments followed traits carefully selected for discontinuity to permit definite and sharp classification rather than "more-or-less" distinctions. This was a key decision. MENDEL restricted his attention to individual traits for each cross, avoiding the "noise" of extraneous characters. His analysis was quantitative and he displayed a mathematical sense in the analysis of data and the design of experiments. He had a clear feeling for probability and he was not put off by large deviations in small samples. MENDEL was a meticulous researcher. The sheer mass of his data is impressive and his experiments built up from the simple to the complex.

The immediate impact of MENDEL's paper, presented in 1866, was nil, although it was distributed to about 120 libraries throughout the world through the exchange list of the Brünn Society and was available in England and the United States. The paper was listed in the Royal Society Catalogue of Scientific Papers for 1866 (England) and even referred to without comment in a paper on beans by H. HOFFMANN in 1869. The only substantial reference was in the 590-page treatise on plant hybrids *(Die Pflanzen-Mischlinge, ein Beitrag zur Biologie der Gewächse)* by the American W. O. HOFFMANN in 1881. He also coined the word "xenia." MENDEL's name is mentioned seventeen times, but it is clear that FOCKE did not understand him. In the critical passage he wrote:

> MENDEL's numerous crossings gave results which were quite similar to those of KNIGHT but MENDEL believed he found constant numerical relationships between the types of the crosses. (In *Die Pflanzen-Mischlinge, ein Beitrag zur Biologie der Gewächse,* Verl. Gebr. Bornträger, Berlin, p. 334, 1881.)

Yet, the period from 1866 to 1900, the classical period of cytology, the study of cells, was to establish the basic part of structural cell biology that put MENDEL's theoretical discovery of inferred genes into structures contained in each living cell. In 1866, the German E. HAECKEL (1834-1919) published his conclusion that the cell nucleus was responsible for heredity. Soon thereafter, the *chromosomes,* the physical framework for inheritance, became the focus of attention in mitosis, meiosis, and fertilization with speculation on its relation to heredity. Later, W. ROUX (1850-1924) postulated that each chromosome carried different hereditary determinants. Nevertheless, the experimental evidence was not reported until 1902, when T. BOVERI (1862-1915) announced that each of the thirty-six chromosomes of the sea urchin were necessary for normal development. The issue was cloudy because the details of the meiotic process were not well understood.

In 1891, the German plant breeder W. RIMPAU published a paper in which he reported about spontaneous and induced hybrids, mainly in wheat and pea. He did not know of MENDEL's paper, but knew of the book of W. O. FOCKE where MENDEL's pea crosses were cited by page. RIMPAU described all his hybrids with intermediate characters considering the parental habits. However, he noticed the great heterogeneity in the hybrid progeny. He found complete setbacks toward the parents, combinations, several intermediate forms, and sometimes completely new traits not seen among the parents.

A student of >>> A. S. WILSON, >>> W. S. SUTTON (England), soon after, recognized in a 1902 paper that the association of paternal and material chromosomes is arranged in pairs and their subsequent separation during meiosis constituted the physical basis of Mendelian genetics. SUTTON wrote two of the most important papers in cytology but never received his doctorate. At the end of 19th century, one American and two Swedish breeders came very close to the discovery of inheritance of single traits. A. ÅKERMAN and J. MACKEY (1948) mentioned that P. BOLIN and H. TEDIN reported in 1897, during the Agricultural Congress in Stockholm (Sweden), about crossing experiments with barley, pea, and vetch cultivars started in 1890. They found no variation in F1 but different combinations of parental features in F2, which could be numerically predicted. In 1897, BOLIN reported in the 2nd Agricultural Congress:

> Concerning the inheritance of typical characters in the second and following generations seems to exist a regularity. The forms that occur namely represent all sort of combinations of parental traits and thus can be calculated with mathematical precision. (In *Nagra Iakttagelser Ofver Vissa Karaktarers Olika Nedarfningsformaga Vid Hybridisiering Hos Korn,* Sveriges Utsadesforen. Tidskr. 7, p. 139.)

In 1901, W. J. SPILLMAN reported during the 15th Annual Meeting of the Convent of American Agricultural Colleges and Experimental Stations about crosses between winter and spring wheat that he began in 1899. As with BOLIN, he wrote:

> When these results were classified, they conformed the above suggestion; and if similar results are shown to follow the crossing of other groups of wheat, it seems entirely possible to predict, in the main, what types will result from crossing any two established varieties, and approximately the proportion of each type that will appear in the second generation. (In *Proceedings of 15th Annual Convent of the Association of American Agricultural College Experiment Stations,* Washington, DC, Official Experiment Station Bulletin 115, 1901, p. 86.)

SPILLMAN also found that the most frequent character in F2 corresponds to the F1 plant, and he knew that quantitative investigations are the key for the understanding of inheritance of single traits. H. F. ROBERTS (1929) reassembled SPILLMAN's experimental data and came to the conclusion that the traits "length of spike," "awnless spike," "hirsute glumes," "glume color," etc., segregate in a 1:2:1 or 3:1 manner.

In a book published in 1885 by NÄGELI and PETER titled *Die Hieracien Mitteleuropas,* MENDEL's paper is cited out of context, lumped with his other paper on inheritance in the *Hieracium* species. The relation between MENDEL and NÄGELI is another shameful episode in the history of academic science. NÄGELI (1817-1891) corresponded with MENDEL and received MENDEL's reprints, a reformulated explanation and packets of seed of peas with notes by MENDEL, but he could not or would not understand the paper. His eternal punishment is that he may only be remembered for this fact. There are other curious references. The Russian botanist I. F. SCHMALHAUSEN (1849-1894) appeared to have read and appreciated MENDEL's results. He is cited as a footnote to a literature review of 1874.

MENDEL received forty reprints of his published paper, of which three have been traced. One went to C. NÄGELI at München (Germany), one to A. K. von MARILAUN at Innsbruck (found after his death, uncut), and one turned up in the hands of M. W. BEIJERINCK, who sent it to >>> H. de VRIES, undoubtedly the true source of VRIES' introduction to MENDEL. Thus, he knew of MENDEL's paper before VRIES. He even claimed to be the rediscoverer of MENDEL's work, at least five years before VRIES (see EMDEN, 1940). By the way, M. W. BEIJERINCK was the first, in 1888, to isolate the bacteria causing root nodulation in legumes. He called them *"Bacterium radicicola."*

In this place should be remembered another man who was also close to the rediscovery of MENDEL's law of inheritance—W. BATESON. >>> BATESON stated in his paper of 1899 during an international conference of the Royal Horticultural Society, London:

> What we first require is to know what happens when a variety is crossed with its nearest allies. If the result is to have a scientific value, it is almost absolutely necessary that the offspring of such crossing should then be examined statistically. It must be recorded how many of the offspring resembled each parent, and how many showed characters intermediate between those of the parents. If the parents differ in several characters, the offspring must be examined statistically, and marshaled, as it is called, in respect to each of those characters separately. . . . (In Hybridization and cross-breeding as a method of scientific investigation, *Journal of the Royal Horticultural Society,* London 24, p. 63, 1899.)

He therefore made the same demand as MENDEL had thirty years earlier, that a statistical treatment of the progeny and a separation of individual traits are indispensable prerequisites for studies on inheritance. One year later in a report for the Evolution Committee of the Royal Society, he introduced the term "allelomorphe" for the opposite characters of a pair of characters, the term "heterozygote" for nuclei that resemble the opposite "allelomorphes" (later abbreviated to "allele"), and the term "homozygote" for identical "allelomorphes." In 1909, he published *Mendel's Principles of Heredity,* explaining his own experiments and those of his students concerning heritable characters. He translated MENDEL's paper into English and was presumably the most vehement promoter of genetics in England after 1900.

Finally, the pieces of the puzzle, however, quickly fit together only after the independent verification of MENDEL's result by the Dutch botanist H. de VRIES (1848-1935), the German >>> C. CORRENS— a student of C. von NÄGELI (1817-1891), and the Austrian >>> E. von TSCHERMAK-SEYSENEGG.

Yet, none of the rediscoverer's papers was in the class of MENDEL's paper in terms of either analysis or style. VRIES was vague on the role of dominance, and CORRENS was convinced that the law of segregation could not be applied universally. TSCHERMAK's paper reported the 3:1 ratio in the first segregation generation but did not interpret the backcross of the hybrid to the recessive parent as a 1:1 ratio, casting doubt of his complete understanding at that time.

In conclusion, MENDEL's paper is a victory for human intellect, a beacon cutting through the fog of bewilderment and muddled thinking about heredity. The story of its origin, neglect, and so-called rediscovery has become a legend in biology. An obscure monk, working alone in his garden, discovers a great biological phenomenon, but the report is ignored for a third of a century, only to be resurrected simultaneously by three scientists working independently.

The genetic revolution had a rapid impact on plant improvement. Although breeders had unconsciously been using many appropriate procedures via crossing and selection in the 19th century, the emerging science of genetics and, especially, the fusion of Mendelism and quantitative genetics put plant breeding on a firm theoretical basis.

The relation between genetics and post-Mendelian plant breeding is best exemplified by two routine breeding protocols. One is the extraction and recombination of inbreds combined with selection to produce heterozygous but homogeneous hybrids, whereby combinations are first disturbed to complete the final order. The other is backcross breeding, in which individual genes can be extracted and inserted with precision and predictability into new genetic backgrounds. The combination of backcross breeding to improve inbreds and hybrid breeding to capture heterosis is the basis of present day maize and other hybrid crop improvement (see Chapter 3.4).

The success of the new science of plant breeding had a substantial impact on agriculture, horticulture, and forestry. Dramatic successes quickly followed; examples include hybrids and disease resistant crops. A further spectacular example of plant-breeding progress was the development of short-stemmed photoperiod-insensitive wheat and rice, the forerunners of the so-called "Green Revolution," for which >>> N. BORLAUG, a plant breeder with the Center for the Improvement of Maize and Wheat, Mexico (CIMMYT) was to receive the Nobel Prize for Peace in 1970.

3.1. REDISCOVERY OF MENDEL'S LAWS—BEGINNING OF GENETIC RESEARCH

H. de VRIES was always involved and fascinated about questions of the theory of the origin of species and the role mutation played in the evolution of plants. In addition, he carried out crossing experiments with *Oenothera lamarckiana* × *O. brevistylis* in 1895, where he observed the uniformity of his crossing hybrids and the "dominance" of some prevailing characters. The detection of a citation of MENDEL's work in the book of FOCKE (1881) and the study of MENDEL's publication guided VRIES to work with peas. In his first report about the segregation of his pea hybrids, he did not mention MENDEL's name but used the expressions "dominant" and "recessive." In the second, a more precise paper, he confirmed MENDEL's result but concentrated again on his mutation theory. He was convinced that breeding efforts should concentrate in looking for spontaneous variations within population caused by "retrogressive" and "degressive" mutations. When he visited the Swedish breeding station Svalöf in 1901, the wealth of different forms was so overwhelming that he tried to convince the breeders that selection of "elementary units" within populations was the only method needed in plant breeding. Of course, this was not MENDEL's approach. Therefore, in VRIES' textbook *Pflanzenzüchtung* (*Plant Breeding*, 1907) he again did not mention MENDEL's paper. His "mutation theory" fit better into breeder's experiences. It became theoretical basis at the breeding station Svalöf. Later, TSCHERMAK visited Svalöf (1901) and presented to a new wheat breeder, >>> H. NILSSON-EHLE, the idea that only Mendelism could provide "a new, rational basis" for breeding of new constant forms by hybridization. The combination of characters could proceed substantially more surely and simply than before. The ongoing discussion about these different opinions suggests how Svalöf became a testing ground for new ideas of heredity and evolution. However, the practical results achieved by the breeders with spontaneous mutations in all sorts of landraces were disappointing. In 1906, >>> NILSSON-EHLE finally preferred hybridization as the most important contribution to breeding programs of Svalöf. In particular, the winter wheat breeding had run into trouble because all the new promising "elementary units," *off-types* or races that had been selected within populations suffered from one or more serious weaknesses. If hybridization could be used to combine good properties and eliminate deleterious effects, progress could be made (NILSSON-EHLE, 1908). At the same time (1909), the American E. M. EAST proposed the idea of multiple alleles at a locus, which had consequences for several breeding programs.

NILSSON-EHLE used to refer always to the visit of TSCHER-MAK as a main source of inspiration for his steps towards crossing methods. In a letter to TSCHERMAK from Cambridge (England) in August 14, 1909, he strongly criticized the former practices of selecting off-types within population to create better varieties:

> . . . than the spontaneous new form, which we specially tried to select and utilize in winter wheat in the years from 1889 to 1905 did not prove–even with extremely strong efforts–better results. The pedigree books show a terrible useless work in this respect; . . . the opinion that the future breeding work must concentrate itself mainly in crossing work. (In P. RUCKENBAUER, E. von Tschermak-Seysenegg and the Austrian Contribution to Plant Breeding, *Vortr. für Pflanzenzüchtung* 48, p. 40, 2000.)

Nevertheless, VRIES, in 1900, independent of, but simultaneously with, the biologists CORRENS and TSCHERMAK, rediscovered MENDEL's historic paper on principles of heredity. He became particularly known for his mutation theory. New elementary species originated by what VRIES (1901) called "progressive mutation." This corresponded to the creation of a new sort of "pangene" *(pangenesis)*. Within a species there could occur "retrogressive" and "degressive" mutations, which correspond to the modification of existing pangenes. Retrogression meant the disappearance of characters through inactivation of pangenes. Degression meant the reappearance of a character; the hereditary differences behaved differently in hybridization. Progressive mutations led to what he called *"unisexual"* hybrids, whereas the hybrids of retrogressive as well as degressive mutations were "bisexual." Only the latter type of hybrid was subject of MENDEL's law. Unisexual hybrids were constant and nonsegregating. VRIES' speculative explanation for this behavior was that in progressive mutations the new type of pangenes would be unpaired in the hybrid. There is no "antagonist," that is, a modified pangene of the same kind with which it could pair up. Therefore, progressive mutation was VRIES' main interest, not Mendelism. He thought that individual selection was the only method needed in plant breeding. The production of new forms through hybridization was superfluous and mass selection could only create "local races." VRIES even gave currency to modern use of the term "mutation" and was a major inspiration for research on spontaneous change of hereditary factors. His book *Die Mutationstheorie (Mutation Theory)*, which he published in two volumes (VRIES, 1901, 1903), made a greater impact in some ways than the rediscovery of MENDEL's laws.

E. von TSCHERMAK-SEYSENEGG was the second Austrian scientist after MENDEL who again detected the laws of inheritance by studying pea crossings. In addition, he was the first plant breeder who purposely applied the combination of genes as a scientific method to improve the agronomic characters and, therefore, the efficiency of cultivated plants in practical breeding. Considering his impressive scientific work, with about 100 original papers, not only can the manifold range of topics be noted but also the originality of his experiments, observations, and theories.

Among other important discoveries, he observed the phenomenon of *xenia* in seedpods of a single F1 plants when he made extended crosses with peas, whereby seed color and seed shape resemble the difference in parental characters at this early stage after hybridization. Xenia effects can then be used to demonstrate the segregation of the parental characters since the seed grains can be classified in alternative groups in the same year of production. Most sporogenous responses are not apparent after the cross until the hybrid seeds are planted and the characters become visible in the progeny plants only. However, in peas, the difference in the color of the cotyledons (yellow versus green) or the shape of the seeds (round versus wrinkled) of the crossing partners shows its F2 segregation of the progenies on their ripe F1 plants. He planted this segregating seed lot individually to follow their behavior in the F2 generation. These observations and the results of some backcrossing procedures, in which parental characters appeared in a 1:1 segregation scheme when the hybrids were crossed again with their parental types, formed the basis of his DSc thesis in January 1900. In this publication he demonstrated and discussed some of the results of his studies—similar data already achieved half a century before by MENDEL. In March of the same year, VRIES (1900) published two similar papers about pea crossing, and coincidentally, C. CORRENS (1900) launched in the April volume of the same periodical *(Berichte der Deutschen Botanischen Gesellschaft)* a third paper about the same topic, but with special reference to MENDEL's experiments and figures from 1866.

C. E. CORRENS, in 1900, independently but simultaneously rediscovered MENDEL's historic paper. He already knew about some of MENDEL's hawkweed plant experiments from NÄGELI. NÄGELI, however, never talked about MENDEL's pea plant results, so he was initially unaware of MENDEL's laws of heredity. However, by 1900, when CORRENS submitted his own results for publication, the paper was called "Gregor Mendels Regel über das Verhalten der Nachkommenschaft der Bastarde" (G. MENDEL's law concerning the behavior of the progeny of hybrids). CORRENS and VRIES were the ones who most clearly redefined MENDEL's laws. CORRENS was active in genetic research in Germany and was modest enough to never have a problem with scientific creditor recognition. He believed that his other work was more important and that the rediscovery of MENDEL's laws only helped him with his other work. He was supposedly indignant that VRIES did not mention MENDEL in his first printing.

Several opinions have been launched about the so-called "independent rediscovery of the Mendelian laws by CORRENS, TSCHERMAK, and VRIES in 1900." There are still some doubts about these "random publication events" that occurred within a few months in 1900. For example, some of the irregularities in the paper of TSCHERMAK (1900) were criticized in recent years (MONAGHAN and CORCOS, 1986, 1987). Why was MENDEL's paper, despite its clarity and incisiveness, ignored for thirty-five years? The best explanation is that MENDEL was ahead of his time and it took that long for the scientific community to catch up. Remarkably, MENDEL's paper was precytological, and the cytological discoveries that were to provide a physical basis for heredity were published between 1882 and 1903. Obviously, 19th century biology was not ready for MENDEL. Part of the reason is that science then, as is now, was conservative. New ideas are absorbed with difficulty and old ones discarded only reluctantly. One paper is not enough.

Soon after the period of rediscovery, TSCHERMAK became fully aware that these fundamental principles of inheritance should be applied to achieve stable and uniform combinations of different characters of parental genotypes by crossings, individual selection, and separate testing of the progenies in all agricultural crops. So, he advocated the system of *"combination breeding"* instead of the only individual ear selection of phenotypically equal plant types within populations, which was very common in creating improved varieties at that time. Due to his excellent crossing techniques and improved selection management, he started first with cereals. With various crossings of rye and wheat varieties of different origins, he tried to solve one of the most important problems, namely, to combine earliness with high yielding performance, cereal breeders were faced with in the dry areas of lower Austria, Moravia, and western Hungary (TSCHERMAK, 1901, 1906). The later performance of many of his varieties in barley and wheat perfectly showed the possibility of successfully combining even these negatively correlated characters.

The situation in the United States was a little confused in the beginning. In 1902, the International Conference on Plant Breeding and Hybridization met in New York City. Reviewing the conference, W. A. CANNON noted that

> . . . generally speaking, the plant breeders had not taken advantage of the Mendelian theory in their work, and some of them did not know of MENDEL or of his experiments before the Conference. (In Review of Proceeding—International Conference on Plant Breeding and Hybridization, *Torreya,* New York 1, p. 13, 1905.)

C. W. WARD, a carnation grower and one of the commercial breeders in attendance, added:

> I have known nothing of MENDEL's theory or law until the day before yesterday, . . . but what I have heard here regarding MENDEL has awakened an increasing interest in the work of hybridizing and I shall secure his books and read them with the greatest interest, for if there is a fixed rule by which I can produce six inch carnations on four foot stems I certainly wish to learn that rule. (In Improvement of Carnations, *Memoirs of the Horticultural Society of New York,* 1, p. 152, 1904.)

However, the dominant force at the conference was W. BATESON. His lead paper combined a straightforward account of MENDEL's laws with a discussion of their applied, and especially commercial, importance. To the breeders he argued:

> Now when we come to the question of the significance of these things to the breeder and the hybridist, it will be found that the significance is exceedingly great. I am afraid of saying that we have reached a point when the practical man who is doing these things with a definite, economic object or commercial object in view can take the facts and use them for his definite advantage. But we do for the first time get a clear sight of some of the fundamentals of which he will in future work, and it cannot be now very many years, if the investigations go on at the present rate, before the breeder will be in a position not so very different from that in which the chemist is. (In Practical Aspects of the New Discoveries in Heredity, *Memoirs of the Horticultural Society of New York,* 1, p. 3, 1904.)

The breeders learned fast. No organization played a more important role in the dissemination of Mendelism than the U.S. Department of Agriculture (USDA). At the 1902 conference, eleven of the seventy-five participants were employed directly by the USDA. Many more were affiliated with the state agricultural colleges and experiment stations. At the first two meetings of the American Breeders Association (ABA) (1903 and 1905), seventeen of forty-five papers were presented by USDA officials. These included the majority of papers dealing primarily with Mendelism, such as WEBBER's "Explanation of Mendel's Law of Hybrids." The USDA also helped popularize Mendelism through its Graduate School of Agriculture, inaugurated in 1902. Seventy-five students attended, of whom twenty-seven were faculty at agricultural colleges and thirty-one were assistants in the agricultural colleges and experiment stations. Mendelism between 1900 and 1910 was thus an applied science. The rapid development of genetics within an agricultural context, where breeding, selection techniques, hybridization, and even evolutionary issues had been addressed in the late 19th century, endowed Mendelism in the United States with a strongly practical and popular aspect. It also ensured that fundamental problems in genetics would be addressed within institutions oriented to practical ends. The subsequent development of genetic research would often reflect dominant social and economic interests in agriculture.

3.2. SCIENTIFIC PLANT BREEDING WITH THE BEGINNING OF THE 20TH CENTURY

With the industrialization of Europe and the United States toward the end of the 19th century, the development of sciences, medicine, and agriculture paced along. It stimulated new products, new technologies, and cross-innovations that also challenged plant breeding. Yields could be significantly increased by utilization of modern crossing and selection methods after 1900 (Figure 3.1). The German philosopher F. ENGELS (1820-1885) recognized crop improvement and plant breeding as important economy factors in his book *Anteil der Arbeit an der Menschwerdung des Affen (The Part Played by Labor in the Transition from Ape to Man):*

> Through artificial breeding, plants and animals are changed under the hand of man in a manner that they are not to be recognized anymore. (In K. MARX and F. ENGELS, *Werke* Vol. 5, Dietz Verl. Berlin, p. 386, 1982.)

At the end of the 19th century, plant breeding became scientific, complicated, expensive, and difficult. It was no longer possible for amateurs to breed crops. Plant breeding was institutionalized, usually in government-owned institutes or in government-financed universities (Table 3.1). In addition, several new institutions arose. In 1886, B. von NEERGARD founded the Swedish Society of Seed Breeding in Svalöf. In 1888, W. RIMPAU inaugurated a Seed Section in the

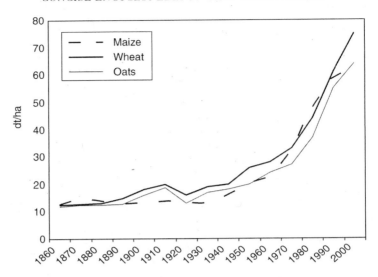

FIGURE 3.1. The development of average yields in grain maize (U.S.), wheat and oat (Germany) over a 100-year period during 1860-1900 almost no progress by utilization of allogamous populations in maize and landraces in wheat and oats; during 1900-1930 slow progress by utilization of cross and combination breeding; during 1940-2000 continuous yield progress by using combination and mutation breeding in wheat and oats; during 1940-1960 significant progress in maize using double crosses; during 1970-2000 tremendous progress in maize by application of single crosses. *Source:* Data compiled from SCHMALZ, 1969, and ALLARD, 1999.

German Agricultural Society. In 1889, K. von RÜMKER held the first lecture on plant breeding at the University of Göttingen (Germany). The first scientific "School of Forestry" was already founded by the University of Jena (Germany) at Zillbach (Thuringia). >>> H. COTTA established it in 1786. In the United States, the first meeting of ABA was held in 1903 and opened by W. HAYS, professor of agriculture at the University of Minnesota and director of the state experiment station.

Breeders always try numerous strategies to capitalize on the insights into heredity. The art of recognizing valuable traits and incorporating them into future generations is very important. Breeders have traditionally scrutinized their fields and traveled to foreign countries, searching for individual plants that exhibit desirable traits. Such traits occasionally arise spontaneously through a process of mutation, but

TABLE 3.1. Important breeding institutions for crop improvement founded from 1800 to 1940.

Institution (location)	Country	Year founded	Breeders/ People involved
Plant Breeding Station (Gross-Enzersdorf /Vienna)	Austria	1903	E. von TSCHERMAK
Institute of Plant Breeding (Lednice)	Austria/ Lichtenstein	1913	E. von TSCHERMAK
Centre de Recherches agronomiques (Gembloux)	Belgium	1872	
Station de Selection du Boerenbond Belge (Héverlé)		1925	A. G. DUMONT
Maison Vilmorin-Andrieux (Verriéres)	France	1815	M. H. VILMORIN and P. VILMORIN
Institut des Recherches Agronomiques (Noisy-le-Roi)		1921	E. TISSERAND and E. SCHRIBAUX
Centre National de Recherches Agronomiques (Versailles)		1923	E. ROUX
Institute of Forest Studies (Tharandt)	Germany	1811	H. COTTA
Seed Breeding Station (Emersleben/Hadmersleben)		1871	F. HEINE
Bavarian Plant Breeding Institute (Weihenstephan/München)		1902	KRAUS
Institute of Plant Production and Breeding (Alferde/Breslau)		1872	F. W. BERKNER
Plant Breeding Institute (Halle/Saale)		1863	J. KÜHN
Agricultural Experiment Station (Salzmünde)		1865	F. STOHMANN
Seed Association (Schwartau)		1901/1925	R. H. C. CARSTEN
Landessaatzuchtanstalt (Stuttgart/Hohenheim)		1905	C. FRUWIRTH
Plant Breeding Institute (Magyarovar)	Hungary	1909	
Hungarian Royal Lowland Agricultural Institute (Szeged)		1924	
Institut di Genetica per la Cerealicoltura (Roma)	Italy	1919	N. STRAMPELLI
Stazione Sperimentale di Granicoltura (Rieti)		1907	N. STRAMPELLI
Institut di Allevamento per la Cerealicoltura (Bologna)		1920	F. TODARO

TABLE 3.1 *(continued)*

Institution (location)	Country	Year founded	Breeders/ People involved
Agricultural Experiment Station (Ettelbrück)	Luxembourg	1884	C. ASCHMAN
Plant Breeding Institute (Wageningen)	Netherlands	1876	L. BROEKEMA
Station de Recherches Agronomiques (Groningen)		1889	
Agricultural Experiment Station (Sadilovsk)	Russia	1896	P. A. KOSTYCEV
Bureau of Applied Botany (St. Petersburg)		1894	A. F. BATALIN and R. E. REGEL
Plant Breeding Station (Moscow)		1903	D. L. RUDZINSKI
Agricultural Experiment Station (Kharkov)		1909	V. J. JUREV
College of Agriculture/Dep. of Genetics (Stellenbosch)	South Africa	1898	A. FISCHER
Elsenburg College of Agriculture (Elsenburg)			J.H. NEETHLING
Svalöf Plant Breeding Station (Svalöf)	Sweden	1886	N. H. NILSSON-EHLE
Weibullsholm's Plant Breeding Station (Landskrona)		1870/1886	W. WEIBULL
Rothamsted Experiment Station (Rothamsted)	United Kingdom	1843	J. B. LAWES
Plant Breeding Institute (Cambridge)		1912	R. H. BIFFEN and F. ENGLEDOW
Institute of Forest Breeding and Genetics (Albany)	United States	1925	J. G. EDDY

the natural rate of mutation is too slow and unreliable to produce all the plants that breeders would like to see and that are required by industrial progress.

3.2.1. Breeding by Selection

Practical breeders have for generations worked without knowledge of scientific principles of inheritance. In many cases, they achieved valuable results. The chief method of the progressive breeder has been, and continues to be, selection, or the choice, of those plants that approach most closely to the type desired. Such selection has been based almost wholly on the appearance of single individuals or group of individuals, without regard to ancestry, progeny, or degree of relationship of the individuals selected. MONS (Belgium), KNIGHT (United Kingdom), and COOPER (United States) all wrote in the 18th and 19th centuries of improvements achieved by selecting superior types from their landraces. In 1843, Sir J. le COUTEUR (1794-1875), a farmer of the Isle of Jersey, noted the diversity of types in his wheat fields. Professor LAGASCA, from the University of Madrid (Spain), visited COUTEUR and pointed out numerous differences in plant types occurring in his wheat field. Selections were made and the progenies tested. Some proved superior to the commercial variety and were of more uniform habit of growth; other selections were of little value. The variety "Bellevue de Talevera," one of these selections, was of commercial importance for many years. COUTEUR also was apparently the first to set down clearly the value of selecting individual plants in the improvement of autogamous small grains, although P. SHIRREFF had earlier used the same method in inbreeding some extensively grown varieties of wheat and oats. Like COUTEUR, he proceeded on the assumption that these selected single plants would breed true. New varieties produced by these means were grown extensively.

Since the introduction of pedigree breeding method, however, ancestry has become to some extent one of the criteria of selection, although in the general practice of farms and early-specialized seed producers, the old method prevailed.

As a means for changing the character of a population, selection has usually been applied to characters that show quantitative variation. In connection with such characters, it is necessary to recourse, since single recessive traits may be fixed in the population. With variable traits, such as size, shape, quality, or others, in which the extremes do not usually breed true, the breeder is forced to depend on the slower process of selecting individuals, with slight variation in the desired direction. Breeders find that some characters yield readily to selection and that progress is rapid up to a certain point, after which it becomes ineffective. On the other hand, the same characters may show no change under the same method. Similar systems of selection can give different results.

3.2.2. Cross and Combination Breeding

For all combination methods in breeding, the crossing of more or less related genotypes with the aim of production of new combinations of traits is common. The theoretical basis is the application of Mendelian laws. Most crosses are carried out within species, that is, between varieties and strains.

The history of barley breeding in North America shows that clear observations and targeted combinations were already practiced before and after 1900. W. M. HAYS was one of the four pioneers of barley improvement in the United States and Canada. Plot tests were reported from Wisconsin as early as 1871. Many barleys, mostly of European origin, were brought to Wisconsin to be tested in comparison with "Manchuria" and "Oderbrucher."[1] After many years of experience, R. A. MOORE decided that "Oderbrucher" was the barley best suited to his state. He improved it first by elite, or mass, and later by pure-line selections (see the following text). In much of the work he was aided by L. A. STONE. In about 1908, MOORE and STONE released "Wisconsin Pedigree No. 5" and "Wisconsin Pedigree No. 6." They were good varieties and are still being grown. Barleys of hybrid origin were tested in field plots as early as 1904. C. A. ZAVITZ at Guelph (Canada) had much to do with the dissemination and improvement of the "Manchuria" and "Oderbrucher" barleys. His variety "O. A. C. 21" was grown on a large acreage in Canada and is one of the important barleys in North America. >>> C. E. SAUNDERS of Ottawa (Canada) led the way in the production of "hybrid" barleys (by crosses). As early as 1893 he was testing varieties of hybrid origin in field plots. Several varieties were foreign genotypes. "Club Mariout" was brought from Egypt in 1904. The two-rowed barley "Hannchen" is a selection of Svalöf (Sweden). Besides the selections from "Manchuria" and "Oderbrucher," there are four others of importance, such as "Atlas," "Trebi," "Horn," and "Tennessee Winter." "Atlas" is a selection of the old "Coast" variety. It is similar to "Coast" but is rather lighter in color than most strains of the original Spanish barley. "Trebi" is a selection from mixed barley imported from the hill country south of the Black Sea. It was widely grown in the Rocky Mountain region and in the prairie states of the northern United States and adjacent Canada. "Horn," a two-rowed selection from European barley, was seeded in Montana and adjacent areas. The awns of most barleys are stiff and heavily toothed. Such barleys were unpleasant to handle. Around 1840 a variety that has a hood in place of the awn was introduced indirectly from Nepal. Its appeal to farmers was immediate and lasting. F. HORSFORD at Charlotte made the first known barley hybrid selection in the United States, in 1879 or 1880. The "Horsford" variety and others originating in the same way have long been marketed under various names, such as "Beardless," "Success," or "Success Beardless." R. WITHECOMBE was among the early breeders to employ "Nepal" in hybrids. One of the more important projects for the production of a hooded barley was started at the Tennessee Agricultural Experiment Station by C. A. MOORES in 1905. One of the most interesting hybrid barley was "Arlington Awnless," developed by B. DERR about 1909. From the progeny of a cross of "Tennessee Winter" on "Black Arabian," DERR isolated an awnless variety. "Lion," a smooth-awned black barley, was another introduction from Russia in 1911. Actually, it was reselected at the Michigan Agricultural Experiment Station from its variety "Michigan Black." Two other examples of successful combination breeding for ecological adaptation in wheat are given in Figure 3.2.

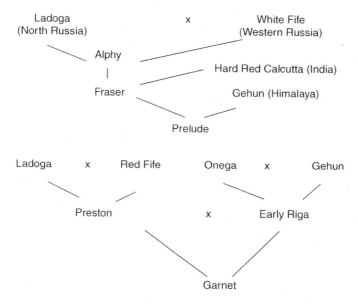

FIGURE 3.2. Breeding for frost-resistant Canadian wheats by combining Russian and Asian landraces. *Source:* Reconstructed after CLARK and BAYLES (1935).

By the imposing development of in vitro techniques during the past five decades, intergeneric crosses also became feasible on a large scale. However, they are mainly used in prebreeding programs. So, crosses can be classified as intervarietal, interspecific, and intergeneric. As long as there is sufficient genetic variability within varieties and strains of species, breeders will prefer this type of intervarietal crosses so as to restrict recombination of the target genotype. For example, when SCHALLER and WIEBE (1952) tried to improve resistance to net blotch disease *(Helminthosporium teres)* in barley, they screened more than 4,500 varieties and strains. Among them 75 were resistant. From them only a few were suitable for the crossing program.

3.2.3. Pure Lines and Improvement of Self-Pollinated Crops

With the beginning of the 20th century, the effect of selection has been carefully investigated. In 1903, the Danish botanist >>> W. L. JOHANNSEN published some results on line selection considering the weights of individual seeds of the "Princess" bean. Because of the autogamous flowering habit of bean, outcrossing could be widely excluded from interpretation of trait variability. He found that plants grown from the lightest and from the heaviest beans of the same mother plant produce seeds of the same average weight, that is, selection among the seeds of same plant was not effective. His explanation of the apparently contradictory behavior was quite simple. There is no hereditary variation, which may be fixed by selection, and selection is, therefore, ineffective. He distinguished two classes of variation:

1. Genotypic variation, or differences in the genotype or hereditary constitution
2. Phenotypic variation, caused by the combined action of many nonheritable factors

Thus, selection is effective only when applied to the first class. He called the group of descendants a "pure line" when the selection among the progeny of a single genetically pure self-fertilized individual was ineffective. At present, the term is more broadly defined as a group of individuals all of whom have the same genetically homo-zygous genotype for one or more loci. The other conclusion of JOHANNSEN's work is that effectiveness of selection depends on the presence of genetic variability (see the following). The pure-line theory has led to a general conception of the way in which selection accomplishes its results. According to this explanation, selection merely sorts out and isolates the genetic factors responsible for the trait selected and does not itself create anything new. Thus, breeding has to both search for and create new variability and selection of the most suitable individuals.

3.2.4. Positive and Negative Mass Selection

The basic way of selection is to choose for breeding those individ-uals that vary in the desired characters, usually by their visible or measurable traits. This is the method of the maize, barley, rye, wheat, or other crop grower who, at harvest time, selects the best spikes from the whole yield and rears the next year's crop from such seed. L. de VILMORIN already recognized the basic importance of separate sowing of individually selected plants. He reported to the Societé Industrielle at Angers (France) in 1856 considering his rapeseed breeding:

> Only when using single individuals, the heritable fitness of plants can be determined. (In *Comptes Rendus des séances hebdomad-aires de l'Académie des Sciences,* Paris 43, p. 172,1859.)

When this principle of selection is kept for many generations, a pe-digree appears, which is the scheme of pedigree breeding (see Chapter 3.2.5). This positive mass selection has been followed by the breeder for many years, and while slow and uncertain, has led to the origin of many improved plant varieties. The opposite way would be to discard all individuals showing negative habit and to propagate the remain-ing, leading to negative mass selection.

The efficiency of positive mass selection was demonstrated in allogamous maize. Already in 1896, selection for high and low pro-tein and high and low oil content of maize kernels were begun at the Illinois Agricultural Experiment Station (United States). As can be determined from Figure 3.3, E. M. EAST and D. F. JONES (1920)

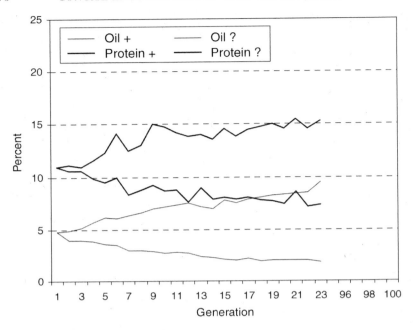

FIGURE 3.3. The results of selection for high and low protein as well as oil content of maize at the Illinois Agricultural Experiment Station from 1896 to 1919 and recent data. *Source:* Data compiled from EAST and JONES, 1920; JANICK, 2004.

were able to demonstrate, after twenty-three generations, that positive and negative mass selections were plainly effective in altering the chemical composition of the kernel. After almost 100 years, in one of the longest experiments ever, U.S. maize breeders have continuously selected to change the oil composition. Typically, mean oil concentration was estimated in about sixty ears, and seeds from only twelve were selected to propagate the next generation. The change after 100 generations in oil concentration was almost continuous and substantial: from a base of about 5 percent, the high-oil-producing line now has about 20 percent oil, and the low-oil-producing line has almost none. Kernel protein concentration gave similar responses, except that the low line reached a plateau at about 5 percent protein. Most changes occurred in the earlier generations (JANICK, 2004).

By the same procedure, new crop cultivars were developed from either primitive wild forms or landraces (Table 3.2). This method is

TABLE 3.2. Examples of crop plant development from wild relatives.

Primitive form	Landrace	Cultivated forms	Advanced varieties
Wild wheat, *Triticum boeoticum*		Wheat, einkorn, *T. monococcum*	
Wild wheat, *Triticum dicoccoides*		Wheat, *emmer*, *T. turgidum*	
Wild wheat, *T. araraticum*		Wheat, Armenian, *T. timopheevi*	
Wild wheat, *Triticum* species	Hulled hexaploid wheat, *T. spelta* *	Hulless wheat, *T. aestivum*, **"Breston"** (1896, U.S.)	**"Mercia"** (U.K.)
Wild barley, *Hordeum spontaneum*	Hulled barley, *H. vulgare*, "Hanna" (1875, Austria)	Hulless barley	**"Weihenstephaner Hilte"** (Germany)
Wild oats from Persia, *Avena sterilis*, *A. fatua* *	Hulled oats, *A. sativa,* "Hopetoun oats" (1864, U.K.)	Hulless oats	**"Tschermak's Frühhafer"** (1931, Austria)
Wild rye, *Secale vavilovii* *	Hulless rye, *S. cereale* "Probsteier" (Austria)*	**"Petkuser Roggen"** (1881, Germany)*	**"Danae"** (1962, Germany)

Advanced varieties derived from primitive forms marked by bold letters; * indicates lineage.

generally effective because most of the characters on which it is employed are apparently dependent on a large number of inherited factors that may be slowly sorted out and accumulated by subsequent selection. For example, the number of wheat varieties grown in the United States increased from 126 in 1919 to more than 500 in 2005. It was mainly achieved by methods mentioned previously. Of course, the selection is ineffective where such genetic differences do not exist or cannot be screened by the mass method. The most rapid results accomplished by selection are those obtained in the differentiation of pure lines from mixed population of self-fertile plants.

3.2.5. Pedigree Selection

Before the rediscovery of Mendelian laws, several crop varieties were developed by this approach. During the period from 1857 to 1874, the English wheat breeder Major HALLET, from Brighton, developed the method of pedigree selection in wheat, oats, and barley, apparently believing that acquired characters were inherited and that improvement induced by favorable growing conditions would be transmitted to the progeny. He raised his plants, therefore, under the most favorable cultural conditions, selected the best seed on the best-developed spike of the more vigorous plants, replanted, and followed the same plan of selection in subsequent years. In this way the wheat varieties "Golden Drop" and "Victoria White," the barley "Chevallier barley," and the oats "Pedigree White Canadian" and "Pedigree Black Tatarian" were produced. TAYLOR, a farmer from Yorkshire (United Kingdom), around 1860, found in a stand of "Victoria White" a mutant showing strong and short straw and a shared spike. It was the origin of the famous "Squarehead Wheat," which came via Denmark (1874) to Germany (called "Dickkopf-Weizen"). One of the first advanced winter wheat varieties, "Rimpau's früher Bastard" (Germany), was based on the same method. It derived from the cross-combination "Früher amerikanischer Landweizen" × "Squarehead" from England and was marketed in 1888.

Pedigree selection by utilizing the knowledge of early scientific genetics is known as the progeny test. In adopting the progeny test, breeders have recognized that permanent improvement must be improvement of the genotype, and the problem was, therefore, to develop some methods of estimating the genotype itself. The best evidence as to the genotypic constitution of an individual is obtained from a study of its progeny. Among plants, seeds from single individuals are separately selected and the progenies compared. The most uniformly satisfactory progenies are saved for further selection and the others are abandoned.

In barley, the nestor of the Moravian plant breeders, E. von PROSKOWETZ, tried without success since 1875 to improve the old landrace of barley "Hanna" by single spike selection at his agricultural estate Kvasice and Tlumacova. In 1904, TSCHERMAK convinced him to change to individual selection and progeny testing, and he could find particularly early and high yielding lines of malting barley. The so-called "Kwassitzer Original Hanna Pedigree" became the mother of a wide range of malting barleys in Europe because of its high grain quality, earliness, yield stability, and high adaptability. In Germany, many years after the release of this variety, thirteen sister varieties ("Mittlauer Hanna," "Eglfinger Hado," "Mahndorfer Hanna," "Heines Hanna," "Oppiner Hanna," "Weihenstephaner Hilte," "Braunes Hanna,"

"Dippes Hanna," "Selchower Landgerste," "Heines Haisa," "Crie-
wener 403," "Rimpau's Hanna," and "Mettes Hanna") demonstrated
this early selection triumph, which was similarly successful in Sweden ("Svalöf's Hannchenkorn" and "Svalöf's Hannchengerste 2") (see
WUNDERLICH, 1951).

Sometimes the selected strains within the pedigree were called
"genealogical lines," not to be mistaken for "pure lines." In allogamous
plants, it cannot be a genealogical line when open pollination occurs.
These are not true pedigrees but maternal pedigrees without knowledge about the paternal contribution. Therefore, the terms "mother-
pedigree breeding" or "maternal polyandry method" were used, as, for
instance, in sugar beet breeding. In maize, between 1905 and 1907,
WILLIAMS (Ohio) developed the remnant seed-testing plan.

3.2.6. Bulk Selection

When breeders want to avoid multiple single-plant selections and
its separate progeny testing, they have to start selection in later generations. The portion of homozygous individuals in the segregating
cross-population is comparably high. It was the American >>> H. S.
JENNINGS who in 1912 found that selfing reduces heterozygosity
by one-half each generation (JENNINGS, 1918). Usually, from the
F3 generation until the beginning of the selection, cross-progenies
are propagated as mixed populations in bigger field plots. However,
the method of *bulk breeding* was first practiced in 1908 by H.
NILSSON-EHLE (Sweden). In the same year he proposed the multiple-factor explanation for inheritance of color in wheat pericarp. In
1920, he introduced the method to the breeding station Åkarp for improvement of winter hardiness in wheat. Therefore, the method was
sometimes called the "Åkarp method." Later, the German geneticist
>>> E. BAUR strongly promoted the method, because it was easy to
handle even for large numbers of cross-populations. It is particularly
applicable for selection of winter hardiness, disease, and insect resistance and is comparably cheap. With numerous modifications, the
method is used in many crop improvement programs all over the
world. As a disadvantage, breeders have to cope with yearly changes
of environments and with lower selection efficiency. H. V. HARLAN
and M. L. MARTINI (1938) as well as C. A. SUNESON and G. A.

WIEBE (1942) demonstrated that just an accumulation of a certain genotype in a given population does not guarantee good expression of its traits in a population. However, partial bulk systems and composite crosses can minimize such doubts as demonstrated in a long-term experiment with barley (SUNESON, 1956).

3.2.7. Backcross Breeding

It is a system of breeding whereby recurrent backcrosses are made to one of the parents of a hybrid, accompanied by selection for specific characters. In 1922, H. V. HARLAND and M. N. POPE described the backcross breeding technique for small grains. HARLAND also introduced later the mass-pedigree system (see PARTHASARATHY and RAJAN, 1953). It is a modification of the bulk method introduced by the Canadian J. B. HARRINGTON (1937).

By backcrossing, the characters of the donor parent are accumulated and the genes of the recipient are superseded. It is often applied when a single character has to be transferred to, for example, a high yielding and adapted recurrent parent (Figure 3.4). The desired character of the donor and the genotype of the recurrent parent must be recovered in the segregating populations. The advantages are the stepwise improvement of the cultivar, the handling of multiple generations per year, and small population sizes. Extensive yield tests are not necessary. Moreover, the results of backcrossing are predictable and need little testing efforts. However, the method is ineffective for traits with low heritability. Another disadvantage is the restricted chance of recombination.

Although the method can be classified as conservative, this is the method for producing genetic engineered crop varieties nowadays. The molecular engineered gene and/or trait will be introduced either in a suitable recipient genotype or in a high-yielding and highly adapted variety and then backcrossed when needed. When parallel backcrosses are used for variety improvement, it is sometimes called *"convergence breeding."* It was preferentially applied in resistance breeding. However, it was also successfully used for breeding *monocarpic* sugar beet. In 1940, first monocarpic sugar beets were described by M. G. BORDONOS from All-Union Scientific Research Institute of Sugar Beet at Kiev (Ukraine). As an interesting reminis-

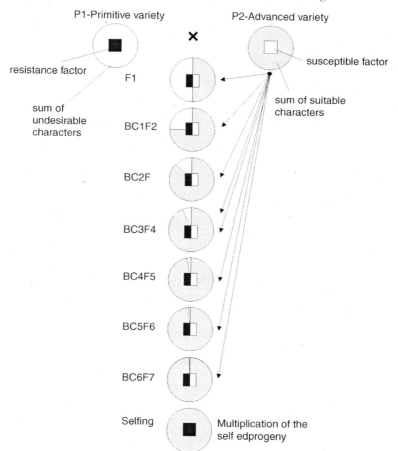

FIGURE 3.4. General scheme of sixfold backcrossing of a disease-resistant landrace to an advanced variety susceptible to disease for improvement of the advanced material.

cence, it should be added that, in 1794, KÖLREUTER already proposed backcrosses, however, with the aim to "change species" of his tobacco hybrids (see Chapter 2.7).

3.2.8. Single-Seed Descent

The motivation behind the *single-seed descent* (SSD) is that, since in many cases early-generations selection can be misleading, the simplest approach is to inbreed the lines in (as much as possible) the absence of selection. One way to do this is to propagate a single seed from each selected plant. This is done under space planted conditions (to reduce competition) until the F5 or F6 generation, at which point the resulting lines are expanded and yield tested. SSD methods (single-seed, single-hill, and multiple-seed) are easy ways to maintain populations during inbreeding. Natural selection cannot influence the population, unless genotypes differ in their ability to produce viable seeds. The artificial selection is based on the phenotype of individual plants, not on the progeny performance. It was first proposed by >>> C. H. GOULDEN (1939, Scotland) and modified by BRIM (1966). It guarantees rapid generation advance by the handling of a large number of crosses and speeds up homozygosity within the population. The SSD method requires little field space for selection; however, additional greenhouse capacity in case of winter annual crops is required. The method is particularly suitable for selecting characters with low heritability.

3.2.9. Near-Isogenic Lines As a Breeding Tool

Near-isogenic lines (NILs) are not fully isogenic. For example, in maize, two distinct composites of F3 lines from a single cross, one consisting of lines homozygous recessive and the other consisting of lines homozygous dominant for a certain gene (with same genetic background), differ only in being homozygous dominant versus recessive for the genes. In wheat, NILs were produced for different *Rht* (reduced height) genes/alleles causing different straw length. The procedure involves the generation of a series of NILs by sequentially replacing segments of an elite line (the recipient genome that is targeted for improvement) with corresponding segments from the donor genome. The objective is to generate a set of NILs containing, collectively, the complete genome of the donor source, with each NIL containing a different chromosomal segment from the donor. In modern breeding, marker-facilitated backcrossing, followed by marker-facilitated selfing to fix introgressed segments, is used to monitor the transfer of the targeted segments from the donor and to recover the recipient genotype in the remainder of the genome. The number of backcrosses required would depend on the number of evaluations that can be made in the marker laboratory. As few as two backcross generations and one selfing generation will suffice if the laboratory resources are adequate to handle the required number of plant samples.

In maize, the NILs are crossed to appropriate testers to create hybrid testcross progeny that are evaluated in replicated field trials (with appropriate checks) for the desired traits. The NILs would be tested per se for crops, such as soybean and wheat. The superior performing testcrosses will be presumed to have received donor segments that contain favorable quantitative trait loci (QTLs). Thus, QTLs are mapped by function, which should be an excellent criterion for QTL detection. The breeding scheme, not only creates enhanced elite lines that are essentially identical to the original line, but it also provides for the identification and mapping of QTLs as a fringe benefit with no additional cost. The scheme is based on having a reasonably good genetic marker map with distinct alternate alleles in the donor and recipient lines.

This breeding strategy can be an excellent procedure for tapping into the potential of exotic germplasm. BROWN, MUNDAY, and ORAM (1989) used a somewhat similar marker-facilitated backcrossing scheme to transfer isozyme-marked segments from wild barley *(Hordeum spontaneum)* into an elite barley *(H. vulgare)* cultivar. Each of the eighty-four NILs was then made homozygous for a single isozyme-marked segment with two selfing generations. After evaluating these lines, per se, in the field, they also concluded that this was a useful approach for identifying QTLs for improving yield in divergent germplasm.

A major advantage of this NIL approach is that once a favorable QTL has been identified it is already fixed in the elite recipient line and the breeding work is essentially completed. Because only a small segment of the genome of the recipient line has been modified, the enhanced line is nearly identical to the original line and the amount of field testing required is minimal. In addition, lines with favorable QTL alleles can be easily maintained and then used for pyramiding several favorable QTL alleles into a single line. A possible disadvantage of this approach is that favorable epistatic complexes between QTLs may not be identified. However, there is little experimental evidence documenting the occurrence of such epistatic interactions.

3.2.10. Polycross Method

It is generally impractical to make large numbers of diallel combinations or to obtain sufficient seed from such matings for adequate
tests. For this, the polycross method was evolved independently by
FRANDSEN (1940) in Denmark, by TYSDAL, KIESSELBACH,
and WESTOVER (1942) in the United States, and by WELLEN-
SIEK (1947) in the Netherlands. TYSDAL and his colleagues proposed the term "polycross" for the progeny from the seed of a line
that was subject to outcrossing with other selected lines growing in
the same nursery. Selected lines or clones are planted in a field isolated from other material of the same crop plant and with several replications of the clones or seed progenies to be tested. By clonally
dividing the provisionally selected plants and placing them throughout an isolated crossing block, one achieved a more nearly random
pollination. Then, if the seed from the various plants representing a
clone in the crossing nursery is bulked, an adequate supply for testing
is available and represents a more nearly random sample of the various possible cross combinations than that from individual clones,
which are pollinated largely by their neighbors. The polycross seed
of each clone is grown out in replicated nurseries at one or more locations, and after the progenies are evaluated for their general characteristics, the general combining ability of the clones may be determined. In the FRANDSEN technique, grass clones were divided into
100 cuttings, and 20 plots of 5 cuttings each are randomly placed in
the various replications of the polycross nursery. The method is particularly suitable for cross-fertilized crops sensitive to inbreeding.

3.2.11. Shuttle Breeding

In 1972, N. BORLAUG developed high-yielding dwarf strains of
wheat while working at the ROCKEFELLER-financed CIMMYT
(Agricultural Station in Mexico City, Mexico). Along with improvements in rice productivity at the IRRI (International Rice Research
Center) at Los Banos (Philippines) and in other crops at yet more agricultural stations, the so-called Green Revolution came into being.
During that time a program was developed by BORLAUG that shut-
tles seed between two (or more) locations to be grown at each. It is
now a proven contributor to success of the CIMMYT breeding pro-

grams (CURTIS, RAJARAM, and MACPHERSON, 2002). Shuttle breeding allows two breeding cycles per year instead of one, for example, a winter cycle in the northern desert of Sonora and a summer crop in the central Mexican highlands. This not only fast-forwards selection, but also exposes test varieties to radically different day lengths, temperatures, altitudes, and diseases. Resulting plants are broadly adapted. They grow well in numerous environments. The method is broadly used in the recent breeding programs of internationally operating companies. They provide the best prerequisites for this sort of seed exchange and seed testing under different environments.

3.3. RESISTANCE BREEDING

The annual economic loss due to plant diseases and pests are tremendous, if all cultivated plants are considered throughout the world. At present, the world spends more than 10 billion USD annually on crop protection chemicals. Despite this, preharvest crop losses due to pests and diseases are estimated at 25 percent. In food crops alone, these losses are enough to feed about 1 billion people!

Many breeding programs center on recombination of disease resistance with other important agronomic characters. Commonly, by subsequent domestication of plants, natural resistance genes get lost or new pathogen races are developed under the selection pressure of large-scale crop cultivation and monoculture. Breeders became aware of the problem from the beginning. According to G. H. COONS (1953), the first and fundamental principle of breeding for disease resistance is that, where host and parasite are long associated, the evolutionary-process-resistant forms are developed by natural selection. The principle has been brought into focus on numerous occasions, when parasites have been introduced into new environments where only susceptible host plants were growing.

Breeding for resistance to late blight disease in potato started in 1850. The German J. F. KLOTSCH in Berlin crossed *Solanum tuberosum* with the wild species *S. demissum* to improve the resistance. Breeding for resistance to wilt diseases (e.g., in *Verticillium* species) has been an accepted practice since the pioneering work of the Americans W. A. ORTON with watermelons and cotton in 1900 and that of H. L. BOLLEY (1865-1956) with flax in 1901. In 1899, ORTON already selected resistant cotton plants utilizing natural and artificial infection. In England, >>> BIFFEN (1905) carried out first inheritance studies on disease resistance to yellow rust in wheat. He found that stripe rust resistance was due to a single gene. Although he misinterpreted incomplete dominance, the work initiated a century of resistance gene use for disease control. In 1915, L. R. JONES and J. C. GILMAN released a *Fusarium* resistant cabbage. In 1920, H. K. HAYES, J. H. PARKER, and C. KURTZWEIL successfully transferred stem rust resistance from *Triticum durum* to bread wheat, *T. aestivum,* in the famous cross of "Marquis" × "Iumillo" as was similarly repeated in 1930 by E. S. McFADDEN. Resistance to stem rust was transferred from *T. turgidum* and *T. dicoccum* species into hexaploid wheat, producing the variety "Hope." It was later determined that stem rust resistance in "Hope" was largely controlled by a single gene located on the short arm of chromosome 3B. The resistance gene was named *Sr2* and assumed to have come from *T. turgidum,* though the original tetraploid accession was lost and could not be confirmed to carry the gene. The cultivar "Hope" was used in Mexico during the 1940s as the donor for developing the stem rust resistant wheat cultivar "Yaqui 48." Since then, the *Sr2* gene has been employed widely by CIMMYT's global wheat improvement program. Germplasm exchange from CIMMYT has distributed it to many wheat production regions of the world. The gene has provided durable, broad-spectrum rust resistance effective against all isolates of *P. graminis* worldwide for more than fifty years. Based on its past performance, *Sr2* has been described as one of the most important disease resistance genes deployed in modern plant breeding.

In 1921, H. K. HAYES and E. C. STAKMAN suggested different races of stem rust that had to be considered in resistance breeding. Before 1900, breeders looked at inherited characteristics as expressed on a quantitative basis, that is, genetically inherited characteristics could be present in varying degrees. For example, a susceptible plant crossed with a resistant plant would produce progeny with every degree of susceptibility, ranging from the one extreme of minimum resistance to the other extreme of maximum resistance. However, the recognition of Mendelian laws in 1900 drastically changed the approach. Now breeders looked at inherited characteristics as expressed on a qualitative basis. The recognition of MENDEL's laws marked the beginning of a conflict between the Mendelians and the biometricians on whether genetically inherited characteristics are based on one single gene or on polygenes. After the description of BIFFEN that wheat resistance can be controlled by a single gene (see earlier), suddenly the Mendelian school of breeders discovered that genetically controlled characteristics that were of economic importance indeed exist. An explosion of research followed. As a result, resistance of many plant diseases was shown to be controlled by single genes. This was the type of resistance that is now called "vertical resistance." A differentiation of horizontal and vertical resistance was first introduced by PLANCK (1963). It was also shown that the quantitative characteristics valued by the biometricians did not conflict with genetic laws. Instead of being controlled by single genes, these quantitative characteristics were controlled by many genes. By the time the conflict with the biometricians was resolved, the Mendelians, focusing on single-gene inheritance, were in complete control of plant breeding.

The single-gene resistances have advantages and disadvantages. The advantages are complete protection against the parasite in question and compatibility with breeding for wide climatic adaptation. These characteristics are attractive to large centralized breeding institutes, since these institutes target large areas and thus implant broad adaptability in their breeding material. The main disadvantage of vertical resistance is its temporary nature, since it breaks down to new strains of the parasite population. Other disadvantages include a loss of horizontal resistance while breeding for vertical resistance and the fact that single genes for resistance cannot always be found. Thus, it has appeared impossible to breed for vertical resistance to some species of crop parasites, including many of the insect pests of crops. Moreover, vertical resistance has been misused in agriculture. Breeders have employed vertical resistance in uniform crop varieties in which every host individual in one cultivar has the resistant gene. During the 1960s, many breeders began to doubt the profitability of breeding for vertical resistance. The commercial life of most vertically resistant cultivars was too short to justify the amount of breeding work. This attitude, combined with the development of improved crop protection chemicals, and the investments of chemical industries in breeding, led to a gradual abandonment of resistance breeding in favor of crop protection by chemicals.

Scientific acceptance of horizontal resistance began slowly. Many plant pathologists still doubt the value of horizontal resistance, and some even doubt its very existence. The reasons are that horizontal resistance is often difficult to measure. It is dependent on environmental effects and often partial. Breeding for horizontal resistance can be cheaper than crop protection by chemicals and it spares the environment. Commonly, it is not necessary to find a good source of resistance, as when breeding for vertical resistance. Transgressive segregation within a population of susceptible plants will usually accumulate all the horizontal resistance needed.

In addition, recurrent mass selection can be applied. This means that about a thousand original parents, high-quality modern cultivars and landraces, are cross-pollinated in all combinations. The progeny should total some thousands of individuals that are screened for resistance by being cultivated without any crop protection chemicals. The majority of this early screening population dies, and the parasites do most of the work of screening. The survivors become the parents of the next generation. This process is repeated until enough horizontal resistance is accumulated. Usually, a maximum of about ten to fifteen generations of recurrent mass selection will produce high levels of horizontal resistance to all locally important parasites. The recurrent mass selection must be performed in the area of future cultivation during the season of future cultivation and according to the farming system of future cultivation. This may produce new cultivars that are in balance with the local agro-ecosystem.

In the historical context of breeding, the wide field of plant pathology and resistance breeding can only be stressed. As already noted at the beginning, resistance breeding and approaches that provide a new source of resistant plants against hundreds of fungal diseases and pests are continuous subjects of research and practical breeding work, including the genetic engineering. In 1995, the first maize variety was registered carrying the insecticidal *Bt* gene (*Bt* = abbreviation of bacterium *Bacillus thuringensis*) providing resistance against the European maize borer, *Ostrinia nubilalis*. It was another milestone in breeding of resistant crop plants (see Chapter 4.2.2).

3.4. HYBRID BREEDING

Hybrid breeding is the development of varieties that are the direct product of a cross between two parents. Hybrid varieties are increasingly grown worldwide. In maize about 65 percent of the area is covered with hybrid material and a yield advantage of about 15 percent, in sunflower 60 percent and 50 percent, in rapeseed 50 percent and 12 percent, in sorghum 48 percent and 40 percent, in rice 12 percent and 30 percent, or in rye 5 percent and 20 percent.

The term hybrid breeding is used in a narrow sense to describe breeding strategies utilizing the effects of hybrid vigor, often observed in the F1 hybrids between plants of two varieties, species, or higher order. In 1916, >>> G. H. SHULL introduced the term "heterosis" to replace "heterozygosis" in describing the phenomenon of luxuriance of hybrids or hybrid vigor, i.e., progeny of diverse (inbred) genotypes exhibit greater biomass, speed of development, and fertility than the better of the two parents [see explantions of KÖLREUTER, SPRENGEL, KNIGHT, GÄRTNER, NAUDIN, MENDEL, DARWIN, FOCKE, or WEBER in Chapters. 2.7 and 3.1 as well as of SANBORN (1890) and of JOHNSON (1891)].

In general, the degree of heterosis increases as the "genetic disparity" of the parents becomes greater. A research project on maize was initiated in 1895 by C. B. DAVENPORT (1908) and P. G. HOLDEN, both students of W. J. BEAL [who in 1871 reported 10-15 percent higher yields in maize hybrids from variety crosses, RICHEY (1920)], at the University of Illinois (United States). In 1900, E. M. EAST (1908), the cofounder of the heterosis conception, was appointed at the department. Although still not known generally, the important effects of heterosis were already discovered. However, they required a causal explanation. G. H. SHULL (1908) started his experiments with maize at Cold Spring Harbor (United States) in 1904, without special consideration of heterosis. Independently, EAST began inbreeding and crossing experiments, which were continued until 1911. He published his work on inbreeding in 1908. The segregation of homozygotes and heterozygotes in inbred lines can be calculated by the binomial expression $1 + (2^r - 1)^n$, where r is the number inbred generations and n the number of genes involved. Before EAST, SHAMEL (1905) reported yields of maize lines inbred for three generations and of hybrids made by crossing inbred lines. In the years 1908 to 1910, SHULL realized that the yield of hybrids succeeded not only the parental inbred lines but also the original stock by 22 percent (ratio = 29 : 100 : 122). With the first genetic interpretation of the results, the first step toward the various theories of heterosis were made. SHULL (1948) has summarized them as follows:

1. The entrance of a sperm into a foreign cytoplasmic environment may in some case produce an initial favorable reaction, which may manifest itself in the F1 and not be repeated or not repeated in the same degree in F2 and subsequent generations (SHULL 1912).
2. Protoplasmic heterogeneity may favor increased metabolic activity.
3. Linked dominant favorable factors may confer some advantages in heterozygotes as compared with homozygotes. The dominance theory was first outlined by BRUCE (1910) and KEEBLE and PELLEW (1910) and expanded by JONES (1917) to include linkage. JONES developed the first commercial hybrid of maize. In this hypothesis, heterosis results from cumulative action of many dominant favorable genes contributing to vigor. The biggest heterosis effect should occur when the maximum number of loci with dominant favorable alleles is present.
4. Not the accumulation of favorable dominants, but the rare occurrence of an unfavorable recessive induces heterosis (JONES, 1945).
5. A sensitization of a nonmutated parent gene by its mutated allele, or vice versa, of the nature of an anaphylaxis produces the heterosis effect (CASTLE, 1946).

Although the dominance hypothesis was the most favored theory of heterosis, evidence has accumulated that indicates that it does not take adequate account of various manifestations of heterosis. Including the molecular studies of recent time, heterosis is concluded as a complex phenomenon of dominance actions and regulatory gene allelic interactions (SONG and MESSING, 2003; BIRCHLER, AUGER, and RIDDLE, 2003). It seems to be a matter of optimization of genetic, physiological, and environmental influences that contribute to the final phenomenon of heterosis, particularly when the breeding application is considered. Fixation of heterosis effects in breeding is feasible in several ways:

- If vegetative propagation of the plants is combinable with heterosis
- If apomixis is associated with heterosis effects
- If heterosis is caused by dominant alleles for a given trait, or bases on complementary gene action, segregating in homozygous manner
- If allopolyploid genome constitutions is present
- If both alleles causing the heterosis effect are available as pseudoalleles
- If complex heterozygosity or inversion heterozygosity of chromosomes permanently produces heterozygosity of certain genes
- If heterosis bases on favorable genotype–plasmotype interactions

In nature, additional mechanisms of maintaining heterozygosity are described, such as *heterostyly, protandry, protogyny,* or sexual differentiation. Thus, not all crop plants are suitable for hybrid breeding. The genetic prerequisites, the economical expenditure, and the expected profit must always be balanced against one another. In maize, the hybrid seed production could be highly optimized by mechanical emasculation (detasseling) and stripe cropping of the parental inbreds. Controlled experiments with varietal hybrids were begun at the Michigan Agriculture Experiment Station prior to 1880 and were carried out until the 1920s. In general, those crosses yielded in excess of the parents. Flint[2] and flour types crossed to dent[2] types generally gave the most sizable yield increase. Later the ear-to-row selection was the first method to be put to rather extensive use. The selected ears were grown out in ear rows the next year, and ears were again selected from the highest-yielding rows. Thus, the principle of progeny testing became an integral part of the breeding system, even though only the female parentage was under control and plants in a population were of heterogeneous nature, which was not reproducible.

Self-fertilization by detasseling the monoecious maize became the standard practice of fixing/isolating the genotypes. Together with their utilization in producing hybrid combinations, the basis for hybrid maize industry was formed. In 1926, Pioneer Hi-Bred Corn Co. in Johnston, Iowa, was the first seed company to organize maize (hybrid) breeding. Now Pioneer Hi-Bred International, Inc., is one of the largest seed company in the world.

After the first experimental four-way hybrid "Burr" × "Leaming" (1917), JONES (1920) suggested using double crosses as a practical way of avoiding low-yielding inbred lines as seed parent in commercial seed production. Subsequently, several modifications of hybrid seed production were developed, which now are standard worldwide (see Table 3.3). The inbred lines used in the beginning of hybrid development were extracted from open-pollinated varieties. More pro-

TABLE 3.3. Basic methods of selection adapted from more than 100-years breeding experience.

Selection for varieties				Maintenance selection
Line selection	Selection of populations	Hybrid breeding	Clonal varieties	Single spike selection
Pedigree method	Mass selection	Variety hybrid[1]		Mass selection
	Positive Negative	Top cross hybrid[2]		Positive Negative
Bulk method	Single-plant selection (Illinois method)	Single hybrid[3]		Single-plant selection
Partial bulk	Single-plant selection with progeny testing (ear-to-row or Ohio method)	Modified single hybrid[4]		Single-plant selection with progeny testing
Pedigree trial	Synthetic varieties	Double modified single hybrid[5]		Recurrent selection
Single seed descent (SSD)		Three-way hybrid[6]		
Backcrossing		Modified three-way hybrid[7]		
		Double hybrid[8]		

[1]Population 1 × population 2 or Variety 1 × variety 2.
[2]Line or single cross × variety, A × population or (A × B) × population.
[3]Line 1 × line 2 or A × B.
[4]Sister lines crosses × line, (A' × A) × B.
[5]Sister line crosses × sister line crosses, (A' × A) × (B' × B).
[6]Single hybrid × line; (A × B) × C.
[7]Single hybrid × sister line crosses; (A × B) × (C' × C).
[8]Single cross × single cross; (A × B) × (C × D).

ductive inbreds were derived later from controlled hybrid combinations. Pedigree systems, topcrossing, backcrossing, or convergent crosses (RICHEY, 1920) have been used in improvement of inbred lines. The value of the method, which is equivalent to double backcrossing, is that it furnishes a plan for the improvement of each of two inbred lines that combine well in a single cross without modifying the yielding ability of the single cross. Multiple convergence, a modification of convergent improvement as presented by RICHEY (1946), implies the convergence of several small streams of germplasm as tributaries to one larger stream for the improvement of inbred lines and their use in crosses. HAYES and JOHNSON (1939) and HAYES, RINKE, and TSIANG (1946) have shown that selection for improvement of combining ability is possible. In 1932, JENKINS and BRUNSON used the *topcross method* to give comparative tests for combining ability. On the basis of a comparison of thirty-six widely grown hybrids adapted to central Iowa and released at intervals from 1934 to 1991, DUVICK (1997) reported that the increase in maize grain yield during that time span averaged nearly 74 kg/ha and year (see Figure 3.1). Hybrid comparisons were based on side-by-side trials, so all of the gain could be attributed to genetic improvement. Earlier studies indicated that genetic improvement usually accounted for about one-half of the total yield increase, with the remainder attributed to changes in cultural practices, such as increased rates of mineral fertilizers and the use of herbicides for weed control and pesticides for control of insects and diseases. It is suggested that the increased grain yielding ability of these widely successful hybrids was due primarily to improved tolerance of abiotic and biotic stresses, coupled with the maintenance of the ability to maximize yield per plant under nonstress growing conditions. Hybrid maize has been repeatedly characterized as "the greatest success story of genetics" (DUNN, 1965, p. 125).

In 1931, M. M. RHOADES discovered *cytoplasmic male sterility* (CMS) in maize. In tobacco it was found in 1932 (EAST); it was also found in radish and sunflower in 1968 (OGURA and LECLERCQ) and in rice in 1969 (SHINJYO). In maize, CMS is a maternally inherited trait that makes the plant produce sterile pollen, enabling the production of hybrids and removing the need for detasseling. Similar mutants were then found in many crops, even leading to innovative systems for hybrid seed production. Recently, in rapeseed, nuclear male sterility is created through genetic engineering of anther specific genes (e.g., *Bcp1*). It was shown that *antisense RNA* inactivation of this unique gene leads to nuclear male sterility in *Arabidopsis* and *Brassica*.

Since the 1930s, when maize breeders started the commercial development of double-cross hybrids, hybrid breeding was also followed by the extensive utilization of single-crop hybrids since the 1960s (see Figure 3.1). For example, hybrid rice was a particular breeding contribution to reduce hunger in the world. It represents an extraordinary achievement since wisdom had considered the incorporation of heterosis impractical in self-pollinated crops. It was successfully developed and released after the discovery of male-sterile cytoplasm at Hainan Island (China) in 1970 (LI and YUAN, 2000). The three-line system uses a male sterile or *A* line, a maintainer or *B* line, and a restorer or *R* line. The first hybrid rice combinations were put into commercial production in China in 1976. The area under hybrid rice increased from 2.1 million ha in 1977 to over 32 million ha in 2005. Recently, hybrid rice has a maximum yield of more than 18 t/ha (average 7.5 t/ha) compared with 5.4 t/ha for conventional cultivars, a yield increase of about 30 percent. Hybrid rice technology has now expanded to India and other Asian countries.

3.4.1. Synthetics

SPRAGUE and JENKINS (1943) described synthetic varieties as populations that are derived from more than four lines after open pollination. They can be grown not only as F1 population but also further. S1 lines of high combining ability may be combined into *synthetic varieties* for fringe-area production and as a stopgap improvement measure as proposed by LONNQUIST and McGILL (1956). Because of testcross progeny performance, a number of lines can be selected to be included in the synthetic variety (see Figure 3.5). The synthetic variety is generally formed by compositing equal amounts of seed of all possible single crosses between the selected S1 lines. In subsequent generations, mass selection may be practiced to maintain the synthetic variety. In cereals other than maize, multiline varieties can be superior over pure line varieties as demonstrated by >>> JENSEN (1965), despite their better adaptability against environmental stress and pathogens.

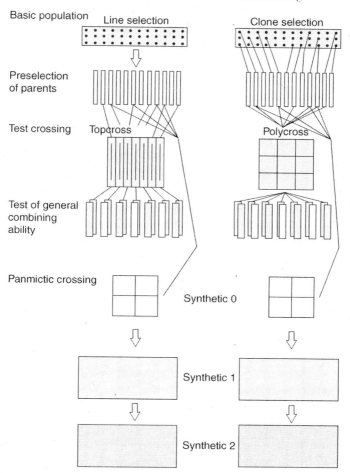

FIGURE 3.5. Breeding schemes for synthetic varieties using inbred lines (left) or clones (right).

3.5. MUTATION BREEDING

3.5.1. Induced Mutation by Mutagens

The carrier of genetic information, the DNA, is stable but not static. Occasionally mutations happen in DNA, changing the sequence of base pairs. A *mutation* can have serious consequences if it changes the synthesis or function of an important protein. However, most mutations have no immediate consequences for the organism, as they take place in noncoding sequences, either within or outside a gene. Such harmless, neutral mutations form the basis for most of the differences between individual genomes.

One of the rediscoverers of MENDEL's laws, VRIES, significantly stimulated the mutation research. In his book *Die Mutationstheorie* (1901), he summarized his observation on spontaneous mutations. Spontaneous mutations of diverse kind have been observed in plants. Nevertheless, there is no evidence that the mutation rate increases because of cultivation, apart from exceptional circumstances, such as unequal crossing-over in hybrids heterozygous for small-scale structural changes in the chromosomes of maize or selection in favor of unstable, "ever-sporting" genotypes in garden pea. The Italian Petrus de CRECENTIA describes the oldest gene mutant in pea in the 13th century. It was a pea with white flowers and white seeds—without doubt—a recessive mutant from which modern vegetable pea varieties derived. Ancient peas found in Hissarlyk (Asia Minor) dated back to 2000 BC, Egyptian findings and peas from the Hallstatt times (750-400 BC) show gray and small seeds and probably purple flowers—a pea type similar to current field peas. Another plant mutant, leaf shape in *Chelidonium majus* (Chapter 2.6), was described by SPRENGER in 1590. The first mutant crop variety seemed to be the so-called emperor rice—a mutant accidentally found by Chinese Emperor KHANG-HI (1662-1723) during a walk along the rice fields. It was characterized by earliness and high quality, and was therefore preferred at the imperial court. For a long time, it was the only variety that could be grown north of the Great Wall. Because of the earliness it sometimes brought two harvests in southern regions.

Famous were NILSSON-EHLE's chlorophyll mutations in wheat and rye (1914), the seed color and fatoid mutations in oats (1907-1921), and the speltoid mutations in wheat (1917-1927). TAMMES (1924) noticed in crosses of wild linseed *(Linum angustifolium)* and cultivated linseed *(L. usitatissimum)* that all important wild and cultivated characteristics of linseed are controlled by the same genes but by different alleles. Thus, the recessive alleles of the cultivated linseed derived their dominant wild alleles by mutations. It means that

changes or improvements from wild to crops can be stepwise produced by single induced mutations.

In coffee, a variety of economically important cultivars based on a
gene mutant with strong *pleiotropic* action, such as "caturra," "mara-
gogipe," "mocca," or "bourbon," arose from primitive forms *(Coffea
arabica* var. *abyssinica).* The spike brittleness of wild barley *(Hordeum spontaneum)* bases on two dominant genes *Bt1* and *Bt2*. Cultivated barleys with the genetic constitution *Bt1 Bt1 bt2 bt2* or *bt1 bt1
Bt2 Bt2* show less tough rachis than *bt1 bt1 bt2 bt2* genotypes. Therefore, the elimination of brittleness in cultivated barley required two
mutations. Many other examples demonstrate the importance of gene
mutations for crop evolution and breeding.

A first comprehensive review on spontaneous and induced mutations was presented from MATSUURA (1933). He also mentioned
that on south Japanese islands a giant radish ("Sakurajima Mammoth
Globe") was bred weighing more than 30 kg, of course, under optimal subtropical growing conditions. N. I. VAVILOV called this radish "the world champion opus of breeding."

Early studies of GAGER (1908), and later by E. PETRY (1922) or
STEIN (1922), failed, however, to demonstrate clearly the *mutagenic
action* of radiation, although effects on chromosomes were described.
Such a demonstration was not forthcoming until the careful experiments by >>> MULLER (1927) on *Drosophila,* STADLER (1928)
on barley and maize, and GOODSPEED and OLSON (1928) on tobacco. These experiments showed beyond doubt that ionizing radiation could greatly increase the mutation frequency of genes, under
certain conditions up to 10,000 times. The so-called *mutation breeding*
accelerated after 1945, at the end of World War II, when the techniques of the nuclear age became widely available. Plants were exposed to gamma rays, protons, neutrons, and alpha and beta particles
to see if these would induce useful mutations. Critical dosages for
dormant seeds ranged between 7,500 r in *Cannabis sativa, Pisum
sativum,* and *Glycine max,* and 50,000 r in *Linum usitatissimum*
(GUSTAFSSON, 1947). Credit for realizing the potentiality of radiation as a tool for plant breeding at an early date must go to two
Russians. In 1931, DELAUNAY published the results of induced mutations in wheat. By 1934 he selected a number of valuable mutations.
SAPEHIN's (1935) results led him to state that artificial mutations

were becoming valuable as a method in plant breeding. Theoretical aspects of utilizing mutations in German plant breeding programs were investigated by >>> STUBBE, SCHICK, KUCKUCK, >>> LEIN, GAUL, >>> FREISLEBEN, and others (>>> W. HOFFMANN, 1951).

After the discovery by OEHLKERS (1943) of the first chemical acting as *mutagen* (ethylorethan) in *Oenothera,* a second wave of mutation research spread through the world. Different chemicals were identified as mutagens, such as sodium azide and ethyl methanesulphonate, dimethyl sulfoxide, methylmethane sulfonate, *N*-Methyl-*N*-nitrosourea, and many others. Examples of crop varieties that were produced via mutation breeding are given in Table 3.4. Despite those mutations that are considered as so-called factorial, gene or *point mutations,* these in a broad sense also include chromosome, genome, and plasmon or plastidome mutations.

Mutation breeding was particularly popular around the world during the 1970s. Although interest has waned somewhat in recent years,

TABLE 3.4. Crop varieties of different genera and species bred by induced mutation programs and application of different mutagens; total number of species in brackets.

Crop type	Species	Number of varieties
Legumes [22]	*Arachis hypogaea*	1
	Cicer arietinum	2
	Glycine max	5
	Phaseolus vulgaris	1
	Sesamum indicum	2
	Vigna mungo	2
	Vigna radiata	9
Cereals [212]	*Hordeum vulgare*	9
	Oryza sativa	3
	Triticum aestivum	195
	Triticum turgidum ssp. *durum*	5
Vegetable [66]	*Lycopersicon esculentum*	13
	Vigna radiata	3
	Capsicum annuum	8
	Brassica pekinensis	4
	Citrullus lanatus	2

TABLE 3.4 *(continued)*

Crop type	Species	Number of varieties
	Cucumis sativus	2
	Raphanus sativus	1
	Spinacia oleracea	1
	Curcuma domestica	2
	Luffa acutangula	1
	Momordica charantia	1
	Solanum melongena	4
	Arctium lappa	4
	Lactuca sativa	6
	Allium cepa	4
	Allium macrostemon	1
	Solanum tuberosum	4
	Ipomoea batatas	5
Oil crops [57]	*Brassica juncea*	7
	Brassica napus	15
	Linum usitatissimum	3
	Helianthus annuus	2
	Sinapis alba	5
	Sesamum indicum	12
	Ricinus communis	4
	Brassica campestris	1
	Cymbopogon winterianus	6
	Olea europaea	1
	Arachis hypogaea	1
Forage crops [42]	*Ornithopus*	
	Lolium	
	Astragalus	
	Medicago	
	Coronilla	
	Alopecurus	
	Festuca	
	Trifolium	
	Lespedeza	
Ornamental plants [556]		
Fruit trees [56]		

Source: Adapted from FAO/IAEA Mutation Variety Database, Vienna, 2004.

occasional varieties continue to be produced using these methods. For example, the new herbicide-resistant wheat variety "Above" (United States) was developed using exposure to sodium azide. Mutation-breeding efforts continue around the world today. Modern mutation research is focused on *site-directed mutagenesis,* i.e., introducing specific base pair mutations into a gene in order to make a protein that differs slightly in its structure from the native type. Computer-aided site-directed mutagenesis is already used in protein engineering. For example, major pollen allergen genes of ryegrass and rice were recently identified and cloned in Australia. These genes are being used to develop new diagnostic reagents and allergy therapeutics by site-directed mutagenesis of adequate proteins. An interesting approach considers the substantial proportion (20-25 percent) of the human population living in temperate and subtropical climates and suffering from allergic rhinitis and seasonal asthma. Pollen grains of grasses including rice are the major cause of hayfever and seasonal allergic asthma.

Of the officially released 2,337 mutation breeding varieties to date, almost 50 percent have been released during the last twenty years. For comprehensive information on mutant varieties the database of the International Atomic Energy Agency (http://www-infocris.iaea. org), Vienna (Austria), is available. This United Nations organization strongly stimulated, over decades, mutation programs for crop improvement worldwide.

A special experiment of utilizing mutants was started during the latter half of the 1920s. Considering the so-called "law of *parallel mutations,*" the meritorious geneticist E. BAUR (1927) was convinced to select a sweet lupin among the wild types. An important prerequisite was a method for rapid screening of bitter alkaloids in single seeds. In 1928, his collaborator >>> R. von SENGBUSCH (1930) was able to screen 1,500,000 individuals; among them he selected three showing low alkaloid content. In 1931, 2 ha were grown from the sweet lupin, and during the next seven years, up to 100,000 ha were grown.

In 1868, C. DARWIN observed that in many tree species belonging to different genera frequently similar growth habits, such as pyramidal or pendulous, could be developed. This apparent phenomenon he called "analogous or parallel variation." He also tried to give a modern explanation:

We can hardly account for the appearance of so many unusual characters by reversion to a single ancient form; but we must believe that all the members of the family have inherited a nearly similar constitution from an early progenitor. Our cereal and many other plants offer similar cases. (In C. DARWIN, *The Variation of Animals and Plants under Domestication,* Vol. 2, 2nd ed., D. Appleton & Co, New York, p. 341, 1883.)

N. I. VAVILOV assumed a general rule behind this. At the end of the 1920s, while he was working on his theory of centers of origin of crop plants (see Chapter 2.2), he also dealt with the "parallel variations" and summarized his ideas in the so-called "law of homologous series," which can be seen as synonymous with "parallel variation."

3.5.2. Somaclonal Variation by In Vitro Culture

Another method for increasing the number of mutations in plants is tissue and/or cell culture (see Chapter 4). Tissue culture is a technique for growing cells, tissues, and whole plants on artificial nutrients under sterile conditions, often in small glass or plastic containers. Tissue culture was not developed with the intention of causing mutations, but the discovery that plant cells and tissues grown in tissue culture would mutate rapidly increased the range of methods available for mutation breeding.

Somaclonal variation was first defined as such by LARKIN and SCOWCROFT, who reviewed the subject in 1981 and were among several authors at that time to draw attention to its potential use for crop improvement. For example, somaclonal variation was exploited in a novel way under a joint program of China and Australia to transfer from *Thinopyrum intermedium* resistance to barley yellow dwarf virus (BYDV) in wheat *(Triticum aestivum).* Rescued single-cell callus cultures from F1 embryos were initiated and induced to form plants showing somaclonal variation, which were then selected for BYDV resistance. Cytological analysis of the genotypes showing stable resistance revealed that chromosomal rearrangement of the chromosome carrying *Thinopyrum* introgressed segment had occurred during the tissue culture phase to confer the stability (BRETTELL et al., 1988).

Later the Biotechnology Center at the Indian Agricultural Research Institutes (IARI) standardized the protocols of plant regeneration of *Brassica carinata* and is now isolating somaclonal variants suitable for Indian conditions. Useful somaclonal variants for earliness, plant height, maturity, etc., have been induced in *B. juncea* and *B. napus*. In dicotyledonous species, EVANS and SHARP (1983) obtained several single gene mutations in tomato regenerated from tissue culture. Somaclonal variation in protoplast-derived plants has been reported in potato (1980), in tobacco (1983), in rice (1985), and tomato (1986).

At the beginning of the 1960s, E. C. COCKING (University of Nottingham, U.K.) used an enzyme preparation for the degradation of cell walls and, thus, yielded single protoplasts from tomato root tips, though it took until 1968 before I. TAKEBE et al. (Institute for Plant Virus Research, Aobacho, Japan) developed a technique for the production of large amounts of active protoplasts from mesophyll cells of *Nicotiana tabacum*, which became standard. J. KYOZUKA, Y. HAYASHI, and K. SHIMAMOTO (1987) established a high-frequency plant regeneration system from rice protoplasts by novel nurse culture methods. "Hatsuyume" was the first rice variety developed by this procedure (SUKEKIYO et al., 1989). Among the progeny derived from "Koshihikari," one R1 line was found to be superior to the original variety in several agronomic traits through R1 and R2 generation and selection. This line was registered as a new rice variety named in 1990. An extensive number of reports soon followed in a wide range of species, indicating that somaclonal variation was widespread and, therefore, accessible to all plant breeders (KARP, 1991, 1995). The amount of references concerning somaclonal mutants and their utilization is large and is growing due to new in vitro approaches. The utilization of new genetic variability has become one of the major objectives of tissue culture. The assembly of genetic variability is vital for improvement of crop plants. Somaclones for the resistance of downy mildew, Fiji disease, and eye spot disease were identified. Application of somaclonal variation in crop improvement can produce increased yield and resistance to biotic as well as abiotic stresses. Some other crops where somaclones were produced are wheat, maize, tomato, geranium, sweet potato, sugarcane, celery, or brown mustard. To date, more than 1,548 new, officially registered varieties of crop plants have been developed through somaclonal variation.

3.6. POLYPLOIDY AND BREEDING

E. STRASBURGER (1910) and H. WINKLER (1916) defined somatic cells and tissues as well as individuals as polyploid if they had three (triploid), four (tetraploid), five (pentaploid), or more complete chromosome sets instead of two as in diploids. WINKLER introduced the term "genome." The state of being polyploid is referred to as polyploidy and may arise spontaneously or be induced experimentally by mitotic poisons. Polyploidy is very common among the crop plants (e.g., wheat, oats, flax, potato, sugarcane, cotton, coffee, banana, alfalfa, peanut, sweet potato, tobacco, plum, loganberry, or strawberry). A. LÖVE (1953) summarized that polyploids are generally more tolerant of climatic extremes at high elevation than diploids, even in the Arctic.

Early in the 20th century, related species with a common basic chromosome number were discovered in a number of plant species. This led to >>> O. WINGE's hypothesis (1917) that polyploid species arose from hybrids between two diploid species, where chromosome pairing either failed in the hybrid or was incomplete. T. H. GOODSPEED and J. CLAUSEN (1928) synthesized an artificial, yet new, fertile hexaploid species *(Nicotiana digluta),* which arose from a cross of the diploid *N. glutinosa* with the tetraploid *N. tabacum.* When breeders learned between the 1920s and the 1950s that several of their subjects were polyploid and often arose either by interspecific crossing or spontaneous doubling of chromosome sets, they recognized the chance of artificial resynthesis of polyploids as in tobacco, in wheat, in *Brassica,* or cotton and induction of new, larger, and vigorous polyploids than the naturally occurring diploids in the same groups. One of the earliest methods of obtaining polyploids was the twin method. In a low frequency of germinating seedlings, twin embryos are occasionally found, which give rise to heteroploid plants. >>> A. MÜNTZING in Sweden (1937) was one of the first investigators to utilize this feature to obtain polyploids. L. F. RANDOLPH in the United States (1932) was able to induce chromosome doubling in maize by applying temperature shocks to germinal tissue.

None of the forgoing procedures was very rewarding from an experimental point of view. It was not until the practical application of the colchicine technique that the way was paved for the production of polyploids in virtually unlimited numbers. It was proposed by A. F. BLAKESLEE and O. T. AVERY (1937) and by B. B. NEBEL (1937). Colchicine inhibits the spindle mechanism at mitosis and results in cells with double or more than double the chromosome number. This was recognized by A. P. DUSTIN in 1934. There are several techniques of application depending on the crop, tissue, cell, or hybrid. *Aneuploids* may occasionally arise after colchicine treatment because of anomalies in chromosome division and separation (see Chapter 2.7).

Autopolyploids were produced in grape plants, rice, rape, einkorn, cotton, beets, buckwheat, clover, alfalfa, timothy gram, melons, apple, carnation, and many others. Doubling of crop plants offers more opportunity for effective recombination. However, selection is more complicated because of the tetra-allelic situation on one gene locus. Genetic stabilization of breeding populations can become extremely difficult. Moreover, the polysomic status leads to chromosomal irregularities during meiosis and mitosis (see Photo 3.1). Aneuploids, seed shriveling, and partial sterility are the consequence. Therefore, vegetatively propagated crop plants are more suitable for autopolyploidization. The two universal effects of chromosomal doubling are increased cell size and decreased fertility. Consequently, crops that benefit most from increased cell size and suffer least from reduced fertility are inherently predisposed to benefit from polyploid breeding. Crops most amenable to improvement through chromosome doubling should:

1. Have a low chromosome number,
2. Be harvested primarily for their vegetative parts, and
3. Be cross-pollinating.

Two other conditions, the perennial habit and vegetative reproduction, have a bearing on the success of polyploid breeding by reducing a crop's dependence on seed production. In the 1950s through the 1970s, induced autopolyploidy in rye was considered an important breeding method. Rye showed good prerequisites for an autopolyploid

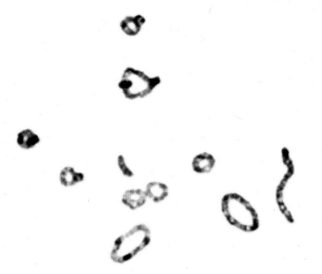

PHOTO 3.1. Meiotic pairing configuration (metaphase I) of chromosomes in autotetraploid rye (Secale cereale, 2n = 4x = 28 chromosomes), showing one univalent, six ring bivalents, one chain trivalent, and three ring quadrivalents. *Source:* R. Schlegel.

crop. Russia, Poland, Germany, and Sweden paid considerable attention to the production of tetraploid rye.

In order to overcome low fertility, seed shriveling, and aneuploid offspring—all three features were believed to be influenced by irregular meiotic chromosome pairing (Photo 3.1)—two major approaches of chromosomal pairing regulation were investigated.

The research group belonging to H. REES at Aberystwyth (United Kingdom) favored an increased quadrivalent formation with convergent or parallel centromere co-orientation as a mean of reduced aneuploidy and, consequently, improved fertility. A "disjunction index" (number of pollen mother cells without univalents and trivalents divided by the total number of pollen mother cells) was used as measure for meiotic pairing regularity. They even demonstrated a positive correlation between the disjunction index and fertility (HAZARIKA and REES, 1967).

Alternatively, J. SYBENGA at Wageningen (the Netherlands, 1964) proposed a preferential bivalent pairing in order to reduce meiotic irregularities by induced allopolyploidization of an autotetraploid. Besides a genetic control of diploid-like chromosome pairing, a complex system of experimentally induced reciprocal translocations was intensively discussed. Despite strong efforts and tests over decades, no practical benefit could be achieved.

R. SCHLEGEL at Hohenthurm/Halle (Germany, 1976) introduced a method for mass production of autotetraploid rye by so-called valence crosses and utilization of gametes of *first division restitution*. Under microplot isolators, nonemasculated tetraploid genotypes (clonal plants) were crossed by spontaneous pollination with diploid genotypes (clonal plants). The tetraploids used as mother plants showed a recessive pale grain character, while the diploids used as male plants showed dominantly green seeds. In this way, green xenia could be selected among the pale grains after harvest of the mother plants. All the green xenia are hybrids, either triploid or tetraploid. Microscopic chromosome counting can differentiate them. The triploids result from fusion of a diploid female gamete and a haploid male gamete, while the tetraploids are derived from a diploid female and an unreduced male gamete (SCHLEGEL and >>> METTIN, 1975). In this way diploid genotypes showing high chiasma frequencies per pollen mother cell (PMC) (>17 chiasmata/PMC) and other useful agronomic characters could be transferred into the so-called half-meiotic tetraploids. The method proved to be efficient for broadening genetic variability of tetraploid breeding material without deleterious effects of colchicine treatment.

To date only a few crop varieties were released showing doubled genomes: tetraploid rye (*Secale cereale,* varieties "Petkuser Tetraroggen," Germany, "Belta," Russia), tetraploid clover (*Trifolium pratense,* varieties "Marino," "Parenta"), tetraploid turnip (*Brassica rapa,* variety "Svalöf's Sirius"), triploid sugar beet, triploid watermelons, triploid banana, and some ornamental plants. While most breeders cross-pollinate plants of a single species, some polyploid breeding methods rely on crosses that can be made between two species within the same genus. A cross between *Musa acuminata* and *Musa balbisiana* produced the bananas with which we are familiar, for example, the triploid banana, variety "Gros Michel." Less commonly, the cross is between members of two different genera.

The successful induction of *allopolyploids* as new crops either by recombining characteristics of separate species or by synthesizing new hybrids may often require more time and effort on the breeder's part. An example is triticale, an allopolyploid hybrid with the genome formula $2n = 8x = 56$ (AABBDDRR) derived from crosses of hexaploid wheat (*Triticum aestivum*, $2n = 6x = 42$, AABBDD) and diploid rye (*Secale cereale*, $2n = 2x = 14$, RR). Its history can be traced back to the end of the 19th century when a first (sterile) hybrid was reported (see Chapter 3). About 100 years later, the first (hexaploid) triticale varieties in which the optimum genome formula was $2x = 6x = 42$ (AABBRR) became successful in agriculture. It is claimed as the first man-made crop plant (see SCHLEGEL, 1996). Induced allopolyploids and autopolyploids were later used for successful transfers of useful characters between species and genera (see Chapter 3.7).

3.7. CHROMOSOME MANIPULATIONS AS A TOOL FOR BREEDING AND RESEARCH

It became clear during the 1930s and 1940s that chromosome behavior and transmission is under genetic control, as are other morphological traits. As >>> M. M. RHOADES (1955) has stated:

> It is evident that genes not only determine the development of the plant as a whole but that the chromosomes themselves, which carry the genes, are also under genic control. (In The cytogenetics of maize. In *Corn and Corn Improvement,* ed. G. G. Sprague, Acad. Press Inc., New York, p. 186, 1955.)

Even the stability of the karyotype is under genetic control in sexual reproduction of plants. This chromosome stability has a profound effect on the regularity of inheritance of characteristics. Irregularities in the transmission of chromosomes introduce biases of concern to the breeder.

In hexaploid wheat *(Triticum aestivum),* it was demonstrated that a dominant gene locus *Ph1* on the long arm of chromosome 5B was responsible for the control of homoeologous chromosome pairing as happens in diploids. It restricts the pairing between homologous chromosomes in a polyploid (RILEY, 1960). Similar mechanisms were later found in polyploid cotton, oats, and polyploid grasses.

When a regular chromosome pairing is given, whole chromosome additions can be established. Particularly in the genus group of *Triticeae,* more than 716 allopolyploid combinations were produced, including species of the genera *Aegilops, Triticum, Secale, Haynaldia, Hordeum,* or *Agropyron* (MAAN and GORDON, 1988).

Autopolyploids and their backcrossing to the diploid parents produces a variety of aneuploid progeny from which trisomic, telo-trisomic, or tetrasomic individuals can be selected. In 1920, A. F. BLAKESLEE, J. BELLING, and M. E. FARNHAM discovered the first trisomics in the weed *Datura stramonium.* In 1921, BLAKES-LEE coined the term "monosomic" for diploid plants missing one chromosome. Because of the cytogenetic instability of trisomics, they do not have a direct breeding value. Their use in genetic analysis and chromosomal localization of important agricultural traits was for many years the only way of gene mapping, particularly in diploid crops. The pioneering work of >>> B. McCLINTOCK—Nobel Prize winner for Physiology in 1983 for the discovery of mobile genetic elements, a discovery that heavily influenced molecular genetics—in utilizing trisomics in maize to associate linkage groups with specific chromosomes is well known (RHOADES, 1955). In 1911, G. N. COLLINS and J. H. KEMPTON demonstrated linkage of genes in maize, and in 1913, >>> A. H. STURTEVANT provided the experimental basis for the concept of linkage versus genetic map distance between loci. By analyzing mating results for fruit flies *(Drosophila melanogaster)* with six different mutant factors, each known to be recessive and X-linked, he constructed the first gene map. He traced each mutation and its normal alternate in relation to each of the other mutants and, thus, calculated the exact percentage of crossing-over between the genes. In 1929, B. McCLINTOCK was the first to report the basic chromosome number 10 in maize ($2n = 2x = 20$). A tremendous impact of trisomic research was also seen in tomato (RICK and BUTLER, 1956), in barley (TSUCHIYA and GUPTA, 1991), and in rye (SCHLEGEL, MELZ, and KORZUN, 1997).

3.7.1. Aneuploids

A cell or plant whose nuclei possess a chromosome number that is greater or smaller by a certain number than the normal chromosome number of that species is called aneuploid. An aneuploid results from nonseparation of one or more pairs of homologous chromosomes during the first meiotic division. This event was described as *nondisjunction* by C. B. BRIDGES (1916). Consequently, monosomic, nullisomic, double-monosomic, trisomic, tetrasomic, double-trisomic, or double-tetrasomic karyotypes may arise.

Before the first sets of aneuploids in allohexaploid wheat were developed in beginning of 1940s by >>> E. R. SEARS, numerous studies were necessary in order to prepare the ground. In 1918, T. SAKAMURA made the first efforts to classify the chromosome numbers in wheat species. In 1922, K. SAX suggested the "A" genome of bread wheat might be related to that in a diploid species, and in 1921, J. PERCIVAL explained the origin of polyploid bread wheat. SEARS was appointed by the USDA as a research geneticist and started his long association with the University of Missouri in 1936 until his retirement in 1980. By both his theoretical and practical contributions, he became the "Father of Wheat Cytogenetics" and stimulated related research in many other crop plants. The discovery of a low level of female fertility in a wheat haploid of the variety "Chinese Spring" and the recovery of aneuploid progeny led to the construction of a vast range of aneuploids that is unequalled in its versatility, practicality, and creativity in any other species known to man. "Chinese Spring" is probably the most famous wheat variety worldwide. It is generally accepted as the standard variety for cytogenetic and molecular research. From this variety, more than 300 aneuploids were developed. Agronomically, however, "Chinese Spring" has some serious faults, such as shattering, susceptibility to almost all wheat diseases and insects, and a poor adaptation to the world's major wheat-growing regions. The answer to why the variety was chosen in which to produce monosomes, trisomes, tetrasomes, compensating nulli-tetrasomes (SEARS, 1965), nullisomics, telosomics, di-telosomics, double-di-telosomics, and almost all of the other possible aneuploids is simple (SEARS, 1959). It came into cytogenetic use by accident. SEARS made hybrids between wheat and rye in 1936 in his attempts to induce chromosome doubling by heat shocks. "Chinese Spring" was used since it was known to cross readily with rye. Among the wheat × rye hybrids, a few wheat haploids were obtained. One of these haploids was pollinated by euploid "Chinese Spring."

Thirteen viable seeds were derived from this backcrossing, showing chromosome numbers of 2n = 41, 42, 43 and reciprocal translocations. Nullisomes were eventually obtained from the monosomes and one nullisome proved to be 3B, which was partially asynaptic and was therefore a good source of additional monosomes and trisomes. It seems that "Chinese Spring" came to the Western world from the Szechuan province of China. British representatives in foreign countries used to be encouraged to collect plants that would be of possible value in their homeland. It was >>> R. BIFFFEN, later director of the Plant Breeding Institute, Cambridge (United Kingdom), who received from Szechuan at about the beginning of the last 19th century a wheat, which he called "Chinese White." This type of wheat was of interest at that time because it was early maturing, set a high number of seeds per spikelet, and was tolerant to drought. In North America, "Chinese White" first appeared in North Dakota, where the pioneering wheat breeder L. R. WALDRON obtained it from R. BIFFEN in 1924. WALDRON shared it with other breeders; one of them passed it on to J. B. HARRINGTON at Saskatoon (Canada). A sample of wheat called "Chinese Spring" came to University of Missouri in 1932 from Saskatoon. It was known at Missouri that this wheat is highly crossable with rye and was therefore acquired by L. J. STADLER. STADLER was very interested in research on polyploidy and amphiploid hybrid production. It seems clear that the "Chinese Spring" was the same as BIFFEN's "Chinese White." BIFFEN's "Chinese White" is still maintained at the John Innes Centre Collection, Norwich (United Kingdom), and is indistinguishable from "Chinese Spring." Recently Y. ZOU (Oregon State University) noticed that a strong resemblance of "Chinese Spring" to certain Chinese varieties is given.

Utilizing "Chinese Spring," SEARS developed the concept that chromosomes from three species contribute their genomes to bread wheat, i.e., chromosome 1 of species A codes for functions that are similar to chromosome 1 of species B, and chromosome 1 of species B codes for functions that are similar to chromosome 1 of species D. This concept of so-called *homoeology* is now fundamental to perception of all allopolyploid species. His joint work with M. OKAMOTO led to deeper understanding of how genes regulate chromosome pairing in polyploid wheat. The recognition of how homoeologous chromosomes are prohibited from pairing and thus immediately allowing a high degree of fertility in a polyploid plant became a cornerstone for modern chromosome manipulations (SEARS and OKAMOTO, 1958) despite the influence on molecular genetics and biotechnology.

3.7.2. Chromosome Additions

Interspecific chromosome addition lines in more or less stable disomic status are mostly known from wheat. K. SHEPHERD and A. ISLAM (1988) counted more than 200. Their cytological stability is insufficient for utilization in breeding but good enough for genetic studies. Extremely difficult to propagate are the wheat–barley addition line 5H or the wheat–*Haynaldia villosa* addition 1Ha and 3Ha. The alien chromosomes are subsequently eliminated on the male and female side. A rarer situation is the preferential transmission of alien chromosomes first demonstrated in the wheat–*Aegilops sharonensis* addition line 4S'. T. E. MILLER, J. HUTCHINSON, and V. CHAPMAN (1982) called this particular chromosome "cuckoo" chromosome. Later it was attempted to recombine this alien chromosome with the standard wheat chromosome 4D in order to stabilize its sexual transmission. Chromosome 4D of wheat carries an important gene *Rht2* for straw length. Because of the spontaneous loss of chromosome 4D in breeding populations, the so-called off-types with longer culms can be identified. Thus, a 4D/4S' recombination might be a way to contribute to more stable wheat varieties (KING, READER, and MILLER, 1988). Other alien chromosome of *Aegilops cylindrica, Ae. triuncialis, Ae. sharonensis,* and *Ae. longissima* were recognized to carry the so-called *gametocidal genes* causing several chromosome aberrations, such as translocations and deletions. They heavily contributed to the establishment of series of wheat chromosome deletions. More than 436 deficient lines with defined chromosomal breakpoints became excellent tools for physical gene mapping for common traits as well as several molecular until recent time (ENDO and GILL, 1996).

Chromosomes added to polyploid wheat are common. Complete sets of wheat–rye (see Photo 3.2), wheat–barley, and wheat–*Agropyron* are intensively used in genetic studies. The most current addition series is the complete set of monosomic and disomic oat–maize

PHOTO 3.2. Different spike morphology of disomic rye chromosome additions to hexaploid wheat var. "Holdfast" (left: control "Holdfast," followed by chromosome additions of diploid rye "King II" 1R, 2R, 3R, 4R, 5R, 6R, and 7R). *Source:* R. Schlegel.

(*Avena sativa*, 2n = 6x = 42 × *Zea mays*, 2n =2x = 20) additions (KYNAST et al., 2001). However, alien chromosome additions in diploid crop plants are still rare. Karyological modifications are less tolerated. The only example published to date is the monosomic and telosomic series of rye–wheat additions (SCHLEGEL, 2005).

3.7.3. Chromosome Substitutions and Translocations

A variation on the wide crossing procedure is to select plants that have single chromosomes or chromosome arms substituted from one species into another. Just for hexaploid wheat, more than 100 different substitution lines are described (SHEPHERD and ISLAM, 1988). Donor species were *Secale cereale, S. montanum, Aegilops umbellulata, Ae. variabilis, Ae. caudata, Ae. comosa, Ae. longissima, Ae. sharonensis, Ae. bicornis, Agropyron elongatum, Ae. intermedium, Haynaldia villosa, Hordeum vulgare, H. chilense, Triticum boeoticum, T. urartu, T. timopheevi*, and others.

After the first reports on spontaneous wheat–rye chromosome substitutions 5R(5A) by KATTERMANN (1937), O'MARA (1947) and RILEY and CHAPMAN (1958), during the past four decades, 1R(1B) substitutions and 1RS.1BL translocations were described in more than 200 cultivars of bread wheat from all over the world (SCHLEGEL, VAHL, and MÜLLER, 1994), demonstrating the enormous impact on wheat breeding (see Photo 3.3). Their most im-

PHOTO 3.3. Metaphase I chromosome constitution with 2 univalents and 20 bivalents in a F1 hybrid of hexaploid wheat from crossing of 1R(1B) substitution variety "Orlando" (Germany) × 1RS.1BL translocation variety "Kavkaz" (Russia) showing both rye chromosome 1R (lower heavy marked univalent) and the wheat-rye translocated chromosome 1RS.1BL (upper univalent with C-bands on the short arm only. *Source:* R. Schlegel.

portant phenotypic deviation from common wheat cultivars has been the so-called wheat–rye resistance, i.e., the presence of wide-range resistance to races of powdery mildew and rusts (BARTOS and BARES, 1971; ZELLER, 1973), which is linked with decreased bread-making quality (ZELLER et al., 1982), good ecological adaptability and yield performance (RAJARAM et al., 1983; SCHLEGEL and MEINEL, 1994), and improved root growth (EHDAIE, WHITKUS, and WAINES, 2003).

The origin of the alien rye chromosome was intensively discussed for genetic and historical reasons. It turned out that four sources exist—two in Germany, one in the United States, and one in Japan. The variety "Salmon" (1RS.1BL) is a representative of the latter and the variety "Amigo" (1RS.1AL) is a representative of the penultimate group, while almost all remaining cultivars can be traced back to one of the two German sources.

There was no doubt so far that the Japanese and the American derivatives differ from each other and from the German sources. Although in two places in Germany—Salzmünde near Halle/Saale (breeder: RIEBESEL) and Weihenstephan near München (breeder: KATTERMANN)—wheat × rye crosses have been independently carried out since the 1920s and 1930s, some authors presumed only one German source (LEIN, 1973).

By cytological and genetic means, it has not been possible so far to verify whether there is a bi- or monophyletic origin of the 1RS.1BL wheat–rye translocation originating from Germany. However, the latest molecular approaches offered a chance to study the identity of that particular wheat–rye translocation occurring in the Weihenstephan and Salzmünde wheat derivatives. A number of DNA probes were considered, which (a) were critical for the short arm of the rye chromosome 1R and (b) should show specificity for the gene pool of Petkus rye. The DNA probe *CDO580* was revealed as the specific one: (1) It clearly differentiated the 1RS.1AL ("Amigo"), 1RS.1BL ("Salmon"), and 1RS.1DL ("Gabo") from the two German sources; and (2) Both the deriving translocation wheats from the Weihenstephan and from the Salzmünde origin showed an identical DNA fragment that was typical for the gene pool of Petkus rye. Therefore, SCHLEGEL and KORZUN (1997) suggested that both German sources have one progenitor in common. Therefore, a single rye chromosome spontaneously introduced into the wheat genome, probably by a single cross, in Germany during the 1930s spread to wheat-breeding programs all over the world, and it can still be seen today.

3.7.4. Chromosome-Mediated Gene Transfer

The early studies on wheat–rye chromosome substitutions showed that almost exclusively dominant genes of the alien species are expressed in the recipient background (ZELLER and HSAM, 1983). Moreover, only a few of the whole-chromosome substitutions remained stable during the breeding cycles. Therefore, from the beginning, there was a breeder's interest to transfer small segments of alien chromatin in order to increase their stable inheritance and to minimize the deleterious effects of donor chromosomes. It was the wheat–rye translocation line "Transec" (KNOTT, 1971) that met these requirements. It was the first induced 4B/2R wheat–rye translocation involving a small piece of rye chromosome 2R carrying resistance genes against leaf rust *(Puccinia recondita)* and mildew *(Erysiphe graminis tritici)* in wheat (DRISCOLL and >>> ANDERSON, 1967). A second example was the induced translocation line "Transfer," carrying a dominant gene for leaf rust resistance of *Aegilops umbellulata* (SEARS, 1961). A third 1AL.1RS translocation was produced in the United States for greenbug resistance *(Schizaphis graminum)* (SEBESTA and WOOD, 1978).

By genetically induced homoeologous recombination, a fourth alien transfer line was established—"Compair" (RILEY, CHAPMAN, and JOHNSON, 1968). It showed the yellow rust resistance of *Aegilops comosa*. Of particular breeding importance was a fifth 4BS.4BL-5RL wheat–rye translocation available in the wheat variety "Viking." The rye segment transferred carries genes for high micronutritional efficiency, such as copper, iron, and zinc efficiency, which have received pronounced attention in recent breeding programs of several of Third World countries (SCHLEGEL et al., 1993). The latter "Viking" translocation and the "Transec" line also revealed that not only homoeologous recombination but also nonhomoeologous recombination could be mediated.

Those experiments of targeted alien gene transfer were only feasible by the high standard of cytogenetic research in wheat starting in 1920s and the specific homoeologous relationships within the *Triticeae*. They brought the first experiences for modern genetic engineering applying molecular techniques. The combination of both approaches resulted in a number of new transfers of resistance to leaf rust, barley yellow dwarf virus, powdery mildew, strip rust, eyespot, or wheat curl mite from *Aegilops, Thinopyrum,* and *Agropyron* species into hexaploid and tetraploid wheat.

Plant breeders will soon change their modus operandi with the development of objective marker-assisted introgression and selection methods. Backcross breeding will be shortened by eliminating the undesired chromosome segments of the donor parent (also known as linkage drags) or by selecting for more chromosome regions of the recurrent parent. Parents of elite crosses may be chosen based on a combination of DNA markers and phenotypic assessment in a selection index, such as best linear unbiased predictors (BERNARDO, 1998). To achieve success in these endeavors, cheap, easy, decentralized, and rapid diagnostic marker procedures are required. The impact of introgression experiments is big. Various studies, mostly conducted on cereals, have estimated that more than 50 percent of the increase in crop production is due to the improvement of crop varieties. It is brought about by transferring desirable genes and traits to crops from landraces and other more distant germplasm sources. Maize hybrids in the tropical and subtropical areas have been improved by 25-40 percent during the past twenty years because of the introgression of a wider germplasm base.

3.8. UTILIZATION OF HAPLOIDS IN BREEDING

3.8.1. Doubled Haploids

Haploidy, i.e., the presence of the half of the common chromosome set, can be a normal as well as abnormal phenomenon in plants, at least in parts of the life cycle. Concerning ploidy mutation, haploids may arise spontaneously. It is necessary to differentiate between monohaploids (deriving from diploids) and polyhaploids (deriving from polyploids). Spontaneous haploids originate either from disturbed fertilization events or from *parthenogenesis.* They were recognized by their strikingly weaker appearance in wheat, maize, potato, tomato, cotton, *Datura,* and *Antirrhinum* crosses with frequencies between 0.01 to 0.5 percent (LINDSTRÖM, 1929; KOSTOFF, 1929; KATAYMA, 1935; HARLAND, 1936; SMITH, 1943). The frequency of spontaneous haploids seemed to be genetically controlled. E. COE (1959) discovered an inbred line of maize showing 3 percent haploids. A. F. BLAKESLEE reported the production of first haploid plants of *Datura stramonium* in 1921 from pollen grains in 1964 by the Indian scientists S. GUHA and S. C. MAHESHWARI.

Already during the 1960s some cytogeneticists pointed to the advantage of haploids for breeding. >>> G. MELCHERS (1960) proposed their utilization in mutation breeding in order to reveal deleterious mutants or to select valuable genes in *hemizygous* allele constitutions. Doubling of haploid individuals would result in direct selection among fully homozygous segregants of cross-populations. The significant reduction of the population size by use of homozygous doubled haploids instead of heterozygous segregants can be taken from Table 3.5.

Double haploids (DH) enable breeders to achieve homozygosity from early generation breeding material of (mainly) self-pollinating crops. The procedure eliminates several generations of selfing nor-

TABLE 3.5. Frequencies and ratios of genotypes in F2 progeny of doubled haploids, diploids, and tetraploids.

Number of alleles	Frequency of recessive plants as determined one of the plants given below (1/x)		
	Doubled haploid	Diploid	Tetraploid
1	2	4	36
2	4	16	1296
3	8	64	46656
4	16	256	1679616
5	32	1024	60466176
n	2^n	2^{2n}	6^{2n}

mally required before uniform lines. It reduces the selection period from thirteen to five years or less. In addition, it provides more accurate and efficient selection of homozygous plants since both dominant and recessive genes are easily expressed in the first generation.

The first criterion for application of DH systems to breeding programs proposed by J. SNAPE et al. (1986) states that DH lines should be produced efficiently from all genotypes. However, the criterion is associated with two problems: the difficulty and efficiency of haploid production through a given DH system and whether varietal differences in haploid production are negligible. Because of the species-specific and genotype-specific differences in production of haploids, three ways of haploid induction can be distinguished:

1. Selection of spontaneous haploids, using suitable (mutant) genotypes and screening methods, e.g., as in maize. Redoubling of haploid seedlings by colchicine.

2. In vitro microspore culture from unripe anthers (anther culture, see Photo 3.4). Microspores are haploid, with n chromosomes. These microspores can be stimulated to chromosome redoubling by introducing specific biochemical treatments and colchicine, thus making it possible—after regeneration—to produce a genetically pure line. This regeneration of DH plants from microspores provides an opportunity for producing fertile and genetically pure (fully homozygous) progeny from one heterozygous parent in a single generation. Several techniques were developed

for a great variety of crop plants, which are successfully utilized in research and breeding.

3. Interspecific and intergeneric crosses and subsequent chromosome elimination.

   3a. The so-called *"Hordeum bulbosum* procedure" is a method for producing zygotic haploids in barley by crossing *Hordeum bulbosum* with *H. vulgare* genotypes. It was first described by >>> KASHA and KAO (1970). After formation of zygotes the *H. bulbosum* chromosomes are subsequently eliminated during embryogenesis, which results in haploid *H. vulgare* plants. After florets pollinated with *H. bulbosum* they are cultured on modified N_6 medium containing 0.5 mg/l kinetin and 1.2 mg/l 2,4-dichlorophenoxyacetic acid. Cultures were maintained at 25°C with a sixteen-hour photoperiod for nine days before embryo rescue. In a comparison of haploid production efficiency using five F1 hybrids from winter × winter and winter × spring barley crosses, 41.6 haploid plants/100 florets pollinated were produced using floret culture. Using detached tiller culture, 13.5 haploid plants/100 florets pollinated were produced. Higher efficiencies achieved with floret culture are attributed to the formation of larger, differentiated embryos. Such embryos lead to higher frequencies of plant regeneration (CHEN and HAYES, 1989). Resulting haploid seedlings are rediploidized by colchicine treatment.

   3b. Wide crosses: Haploids and doubled haploid wheat plants have been produced successfully through wide wheat × maize crosses (and embryo rescue system) since the first report by D. LAURIE and M. BENNETT (1988). Although this system has been found to be less recipient-genotype-dependent and more efficient in haploid production than the *H. bulbosum* system, there is a considerable varietal difference in the efficiency among wheat varieties. It has been suggested that maize genotypes may affect the efficiency of haploid production too. The technology is not anymore restricted to wheat haploidization but also to other cereal species.

The production of doubled haploids has become a necessary tool in the advanced plant breeding of many crop species. In 1980, "Mingo"

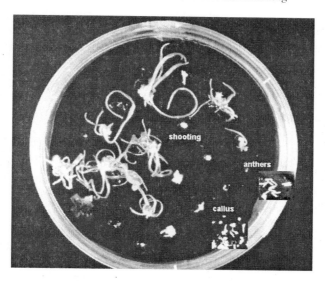

PHOTO 3.4. In vitro culture of wheat anthers for callus formation, haploid plantlet regeneration and doubled haploid production. *Source:* K. Soon-Jong, Suwon, modified.

was the first DH barley released. In 1985, "Florin" was the first released wheat, and "Dellmati" rice was released from Louisiana (United States) in 2003. China, in 1990, released doubled rice haploids (grown on 800,000 ha), and DH wheat (grown on 650,000 ha). Until 2003 more than 230 varieties were registered that derived from DH procedures (Table 3.6).

3.8.2. Dihaploids

Dihaploids of allo- or autotetraploids, that is, a haploid individual containing two haploid chromosome sets, are unique material for genetic research and breeding. They offer the advantage of disomic rather than tetrasomic inheritance patterns and a new approach to gene transfer from numerous wild and cultivated species. In potato (*Solanum tuberosum,* 2n = 4x = 48) they were utilized for several approaches as mentioned before. Potato breeding in the modern sense began in 1807 in England, when >>> T. A. KNIGHT made deliberate hybridizations between different varieties by artificial pollination, and flourished during the second half of the 19th century, when many

TABLE 3.6. Number of doubled-haploid crops commercially utilized.

Crop	Total number of varieties	Important countries/ Number of varieties released
Barley	116	Canada / 36 United States / 3 Australia / 1
Rapeseed	47	Canada / 35
Wheat	21	United Kingdom / 5 Hungary / 5
Melon	9	Spain / 9
Pepper	8	Spain / 8
Rice	8	China / 6
Asparagus	7	Italy / 5
Tobacco	6	China / 6
Eggplant	5	Spain / 5

Source: Adapted from AHLOOWALIA, MALUSZYNSKI, and NICHTERLEIN, 2004.

cultivars were produced by farmers and hobby breeders. The Americans HOUGAS and PELOQUIN (1958) found dihaploids first in the progeny of interspecific crosses *S. tuberosum* × *S. phureja* (2n = 2x = 24). They appeared spontaneously. Decapitation of the seed-parent has resulted in a ten- to fiftteen-fold increase in fruits per pollination and a consequent increase of haploid production. By use of genetically different males, the haploid frequency can again result in about a ten-fold increase. CHASE (1963) proposed an analytical breeding scheme at the dihaploid level followed by resynthesis of the tetraploid.

For forty years those dihaploids heavily contributed to genetic analysis, gene mapping, taxonomic relationships, and, finally, potato breeding. Recent molecular studies in potato would not be feasible without the utilization of dihaploids.

3.9. GRAFTING METHODS

At the beginning *grafting* was the horticultural practice of uniting parts of two plants so that they grow as one. The scion, the part grafted onto the stock or rooted part, may be a single bud or a cutting that has several buds. The stock may be a completely mature plant, such as an apple tree, or it may be a root (usually of a seedling). The most important reason for grafting is to propagate hybrid plants that do not bear seeds or plants that do not grow true from seed. It is also used in dwarfing and in tree surgery to increase the productivity of fruit trees by adding to the number of buds, to adapt a plant to an unfamiliar soil or climate by using the roots of another plant which thrives in that environment, to combat diseases and pests (e.g., the phylloxera) by using a resistant stock, and to promote sexual growth of clones in forest trees. Grafting does not produce new varieties in terms of genetic recombination, since both stock and scion retain their characteristics. In the history of grafting, the transfer of traits of the rootstock to the scion was often a matter of controversial discussion. T. D. LYSENKO,[3] a Russian horticulturist, was one of last advocate of Lamarckianism and the idea of gene transfer via grafting. Together with his political influence in Russia during the 1930s through the 1950s, he misguided Russian genetics and breeding for several years.

Nurserymen of ornamental and vegetable plants or fruit and forest trees mainly use grafting. In general, only closely related plants can be grafted successfully. As a rule, the process is begun when the scion is dormant and the stock is just resuming growth. There are many methods of grafting, all of which depend on the closest possible uniting of the cambium layers of both scion and stock.

The method of transplanting scions was applied since prebiblical and biblical times (see Genesis 2:15; 3:19,23 or 4:2). A cuneiform fragment from Mesopotamia as early as 1800 BC concerns grape bud wood and suggests that the technique of grafting was known at that time. It was also known that the olive tree requires grafting. Ungrafted suckers produce a small worthless fruit. In 58 AD, this explains the power allegory of Paul in Romans:

> For if you have been cut from what is by nature a wild olive tree, and grafted, contrary to nature, into a cultivated olive tree, how much more will these natural branches be grafted back into their own olive tree (Romans 11:24).

Grafting was extensively employed in Roman times. In 323 BC, THEOPHRASTOS of Eresos (371-287) describes six varieties of apples and discusses why budding, grafting, and general tree care are required for optimum production and says seeds usually produce trees of inferior quality fruit. In his *Historia Plantarum* he describes more than 500 plant species. The Indian sugarcane is mentioned for the first time among them. Moreover, he discussed, in detail, which plants can be propagated by fruits and which do not grow when they multiply by seeds, e.g., grape wine, apple, olive, figs, quince, pears, or pomegranate. Almonds derived from seed propagation do not taste good. The latter appraisal shows that grafting seemed to be a common method for clonal propagation in fruit trees during that time. PLINIUS describes in *Historia naturalis* the grafting of quince:

> When the common quince [*Cydonia oblonga*] is grafted on strutea [a greenish wild quince, sparrow apple] one get a separate species, the Mulvian quince [a variety], which is even ate raw as the only one of the genus. (In H. RACKHAM, *Natural History*, Loeb Classical Library, W. Heinemann Ltd., London, 1971, book XVII.)

He described many examples of grafting and emphasized the changes of features on scions including the taste of the fruits after grafting. Grafting of herbaceous vegetable crops was described in the 5th century AD in China. Chinese horticulturists were also the first peony breeders, and their grafting work led to the introduction of many huge, double-flowered trees and herbaceous peonies. It is believed that, during the 12th century, Chinese horticulturist began using grafting techniques to reproduce valued cultivars.

In the 17th century in Korea, a technique to obtain large gourds is described by approach grafting four root systems to a single shoot. The bottle gourd *(Lagenaria siceraria)* was used as a rootstock for watermelon in the 1920s to overcome yield decline from soilborne diseases associated with successive cropping. The development of plastic films in the 1960s led to widespread production of grafted vegetables in Japan and Korea of solanums (tomato, eggplant, and pepper), cucurbits (melons and cucumber), and ornamental cactus. A number of unique advances were achieved including seedless watermelon developed by >>> H. KIHARA (1951) in Japan. The choice of appropriate rootstock permits resistance to soilborne diseases, promotion of scion vigor and yield, and incorporation of stress tolerance. Graft technology of vegetables seedlings has been adopted in Europe.

A special branch of grafting is micrografting: When meristem tip culture fails, it is possible to graft small meristematic domes on young seedlings growing in vitro. G. MOREL and C. MARTIN produced the first successful micrografts in 1958. In the same way, L. NA-VARRO and J. JUAREZ (1977) eradicated all the virus diseases from Spanish citrus orchards. This technique was also very successful in eliminating virus diseases from peach trees (MOSELLA et al., 1980).

3.10. QUANTITATIVE TERMS
IN BREEDING AND GENETICS

3.10.1. Plot Design, Field Equipment,
and Laboratory Testing

By end of the 19th century agronomy as a science had developed from the old style of variety trials, crop rotation fields, and soil culture experiments, when field culture was an empirical art. Research workers in the science of crops and soils and other interested agronomists formed their own circles of discussion, such as the American Society of Agronomy in 1907. In regard CARLETON (1907) stated:

> As a science it investigates anything and everything concerned with the field crop, and this investigation is supposed to be made in a most thorough manner, just as would be done in any other science. Thus, agronomy is the laboratory and workshop of many sciences: agrobiology, chemistry, botany, ecology, genetics, pathology, physics, physiology, and others concerned with the problems of crops and soils. (In Development and Proper Status of Agronomy, *Proc. Amer. Soc. Agron.* 1, p. 18, 1907.)

In Europe, J. B. BOUSEINGAULT established the first experiment station in 1834, being the first to undertake field experiments on a practical scale (see Table 3.1). He farmed land at Bechelbronne, Alsace (France), and investigated the source of nitrogen in plants and systematically weighed the crops and the manure applied. He analyzed both and prepared a balance sheet. Furthermore, he studied the effects on plants when legumes were in the crop rotation. One of his conclusions was that plants obtain most of their nitrogen from the soil. In 1915, J. A. HARRIS proposed a criterion for measuring soil variability that he called a "coefficient of soil heterogeneity." Five years later HARRIS reported that soil heterogeneity is practically universal throughout the world:

> The demonstration that fields upon which the plot tests have been carried out in the past are practically without exception so heterogeneous as to influence profoundly the yields of the plots emphasizes the necessity for greater care in agronomic technique and more extensive use of the statistical method in the analysis of the data of plot trials if they are to be of value in the solution of agricultural problems. (In J. A. Harris, Practical universality of field heterogeneity as a factor influencing plot yields. *J. Agric. Res.*, Washington 19, p. 281, 1920.)

J. B. LAWES established the Rothamsted Experimental Station on his farm in England in 1841. The systematic field experiments began in 1845 and have continued to present day. J. H. GILBERT mainly performed them. For years these experiments have been models of carefully planned trials. In 1859, the Kleinwanzlebener Saatzucht (now KWS) was founded near Magdeburg (Germany); M. RABBE-THGE became the first person in Germany to select beets for seed extraction by ascertaining the specific weight according to sugar content. The first private breeding enterprises emerged in the second half of the 19th century. Many originated at farms and have specialized in specific plant species. Currently there are about fifty companies in Germany with their own breeding programs. The predominantly medium-sized companies have amalgamated to form the Federal Association of German Plant Breeders (Germany). The latter derived from the first Society to Promote Private German Plant Breeding in 1908.

Plant breeding in Sweden was started by farmers. The principal force was B. WELINDER, the owner of the estate Heleneborg in Svalöf. Owing to his initiative and under the chairmanship of Baron F. G. GYLLENKROOK, the South Swedish Association for Cultivation and Improvement of Seed was founded in 1886. Only a few years later, the General Swedish Seed Company was established in 1891.

In 1875, California was the first state to initiate an experiment station, influenced by the Morrill Act signed by A. LINCOLN in 1862. It began field experiments on deep and shallow plowing for cereals. A station was started in North Carolina in 1877, after which many others followed. After the Hatch Act was passed by Congress in 1887, the present USDA stations began to be established. Already twelve were in existence at that time. The Adams Act (1906) provided increased funds. Among the contributions of field-testing in maize have been the discovery that the show-type ear is unrelated to the performance in the field, that ear-to-row breeding may not lead to improvement in yield, and that the combination of inbred lines in hybrids had resulted in higher yields. In 1907, J. P. NORTON substituted the rod-row technique for breeding small grains in place of small plot tests. In wheat, the discovery of rust and bunt resistance and of physiological races of the pathogen has enabled breeding for resistant varieties. At the time "Marquis" wheat was one of the most widely known improved varieties in the United States.

Learning from several inaccessibilities of early experiments, new types of precise field-testing were proposed. Among them long-term experiments were introduced. >>> J. KÜHN began such an experiment with rye in 1878 at the University of Halle/Saale (Germany), which lasted until recent days. In 1911, MERCER and HALL discussed questions about numbers, plot shape and size, and replications in field trials. It remained as a permanent issue in breeding and research. Later >>> R. A. FISHER at Rothamsted (United Kingdom) brought modifications in the field common experiments to make them more amenable to statistical treatment. His basic idea was always to combine experimental planning, field design, and statistical treatment.

One of the simplest experimental designs for testing the yielding capacity of a group of varieties, particularly if the number is not large, is that of randomized blocks. In such designs, the varieties are grown in random order in each of several complete replication series or blocks, the number of replications used depends on the degree of precision desired for the comparisons of the variety means. FISHER stated that randomization of the order of varieties in a block must be followed if an unbiased estimate of error is to be obtained.

Since the 1920s, and influenced by statistical advances, the block method was introduced in experimental field design. Most popular became the so-called "Latin square" or "Latin rectangle." With heterosis breeding and polycross methods the number of treatments increased. For that, the so-called lattice designs were developed. Later polyfactorial trials were developed in order to consider more than one varying factor. Following is a list of *experimental designs:*

- Orthogonal hierarchical designs (randomized blocks, split-plots, or split-split-plots)
- Factorial designs (with blocking)
- Fractional factorial designs (with blocking)
- Lattice squares
- Balanced latin squares
- Semi-Latin squares
- Alpha designs
- Cyclic designs
- Balanced-incomplete-block designs
- Neighbor-balanced designs
- Central composite designs used to study multidimensional response surfaces

In order to cope with the many concerns of breeders, a special field-plot technique was developed over the years. Various types of seeding, harvesting, and threshing equipment were applied and tested. In the 1930s, seeders were described in the United States as having an endless-belt mechanics. It permitted much more accurate seeding. The seed is distributed evenly over the belt distance corresponding to the length of the row to be seeded, and the ratio of the belt speed to ground travel determines the seeding rate. Since the beginning of the 1930s, >>> O. VOGEL from Washington State University, Pullman, designed wheat nursery seeders adapted for one- or three-row seeding and space or drill planting. Space planting is accomplished almost as rapidly as drill planting, and the planter seeds to the last grain (see Photo 3.5).

A four-row cereal-nursery seeder was developed at the South Dakota Experiment Station in 1949. It was a modification of the V-belt principle and the utilization of a small tractor for power. In 1933, as

PHOTO 3.5. Hand sewing of cereal trials 70 years ago (top) and modern pneumatic precision space planter for seeding experimental plots (bottom). *Source:* M. Höller, Wintersteiger AG, Austria.

for small seeds, a two-row press-wheel maize and cotton planter was adapted by J. R. QUINBY and C. STEPHENS, scientists of Texas Agricultural Experiment Station. Plant hoppers were removed and a downspout water pipe used to seed into from the operator's lap. Each planter was equipped for two men to seed and carry the seed packets arranged in boxes.

For harvesting, a rotary-shear cutter for cereal rod rows was established and a first nursery thresher as well as individual-plant thresher was described by KEMP (United States) in 1935. BROWN and THAYER designed a power machine for cutting rod rows and a bundle-tying device in 1936. A mechanical harvester for small-grain nurseries was described by ATKINS in 1953. It was designed for cutting the two center rows of a four-row plot, although it could be used also for cutting single rows. ARMSTRONG and COOPER designed a cyclone-type thresher in 1948. Its principle advantage was the ease of cleaning. A furnace-type fan with rubber facing was mounted and powered through a four-step pulley. LILJEDAHL and his team, in 1951, constructed the first self-propelled combine-harvester for small-grain field plots.

3.10.2. Statistics in Breeding

The major contribution of modern plant breeding, from a genetic point of view, as practiced over the last century, has been the maximization of the expression of traits of agronomic interest, principally the tolerance to biological and environmental stresses (DUVICK and CASSMAN, 1999). This has been achieved primarily through the application of powerful tools in quantitative genetics and statistics (see Table 1.1).

The progress of practical breeding (i.e., selection among big numbers of individuals, handling of experimental plots, and description of numerous characters) required an increasingly precise treatment of quantities. In 1812, C. F. GAUSS (1777-1855), the German mathematician already suggested the theory of least squares. By 1820 he began to explain the idea of "probable error."

According to HACKING (1965), J. ARBUTHNOTT (1710) was the first to publish a test of a statistical hypothesis. HOGBEN (1957) attributes to J. GAVARRET (1840) the earliest use of the probable error as a form of significance test in biology. He also stated that VENN (1888) was one of the earliest users of the terms "test" and "significant." The use of chi-square for testing the goodness of fit for segregation ratios was suggested by J. A. HARRIS in 1912.

F. GALTON, who is accepted as creator of empirical human genetics and who introduced the term "eugenics," applied the twin studies for inheritance of human characters and biostatistic investigations for theoretical and practical genetics. The introduction of mathematical methods in breeding and genetics, particularly, is the merit of L. A. QUETELET (1796-1874). His mathematical treatment of variability founded the English school of biometricians. GALTON (1889) introduced the quantitative measurement in segregating populations. He treated populations as units and found several laws. The law of regression is one of them. K. PEARSON (1900), a student of GALTON and the founder of the journal *Biometrics,* further developed this law and made a contribution to mathematical aspects of testing. He introduced the term *"standard deviation."*

Based on a counterattack against biometricians of the late 1800s who misunderstood MENDEL's law of segregation, W. E. CASTLE (1903) discovered an equilibrium law for not only the 1:2:1 ratio expected in F2 generation but also for an infinitely large class of ratios that could result from selection in any generation. Thus, he had unknowingly discovered the basic rule of population genetics, the HARDY-WEINBERG rule.

In 1908, the English mathematician G. H. HARDY (1877-1947) and, in 1909, the German physician W. R. WEINBERG (1862-1937) independently developed the law of equilibrium for populations. One locus with two alleles at frequencies of p and q will maintain those frequencies over an infinitely large random mating population with no mutation, selection, or gene flow. If p is the frequency of A alleles in the population, and q is the frequency of A' alleles ($p + q = 1$), then the frequency of individuals in the next will be as follows: p^2 (AA individuals), q^2 ($A'A'$ individuals), and $2pq$ (AA' individuals), where $p^2 + 2pq + q^2 = 1$.

The importance of GALTON's "standard deviation" was later proved by the Danish geneticist W. JOHANNSEN in terms of dissociation of populations into pure lines. Experimental field design, sampling, analysis of variance, the theory of measuring observational error, and refining procedures are associated with longstanding director of the Agricultural Research Station Rothamsted, R. A. FISHER. In 1918, he introduced the ideas of quantitative inheritance and correlation:

> In general, the hypothesis of cumulative Mendelian factors . . . the influence of non-genetic causes such as environment, . . . seems to fit the facts observed for quantitative traits very accurately. (In The correlation between relatives on the sup position of Mendelian inheritance, *Transact. Royal Soc.,* Edinburgh 52, p. 412, 1918.)

In 1922, FISHER proposed the analysis of variance and other statistical methods, such as "degrees of freedom" to modify the meaning of chi-squared. It caused a clash between him and K. PEARSON, who proposed it first in 1900. Nevertheless, FISHER's papers[4] became the fundament of quantitative genetics and had great impact on the development in 1950s and 1960s. A promising tool to bridge the gap between Mendelian and quantitative genetics came with the development of methods for mapping of quantitative trait loci during the 1990s.

The basic idea of quantitative treatment of field experimentation traces back to K. SAX (1923), to S. WRIGHT's (1921) "Biometrical relations between parent and offspring," and to R. A. FISHER *Statistical Methods for Research Worker* in 1925. However, according to E. S. BEAVEN (1935), T. B. WOOD and STRATTON were the first to determine probable errors in the context of replicated agricultural experiments. Yet the foundations of modern hypothesis testing were laid by FISHER (1925), although the modifications propounded by NEYMAN and PEARSON (1933) are the generally accepted norm.

One of the essential test methods applied, the so-called "t-test," was introduced under W. GOSSET's (1876-1937) pseudonym "Student" (therefore, "Student's test"). He established tables for estimating the probability that the mean of a unique sample of observations lies between "minus"/"infinity" and any given distance of the mean of the population from which the sample is drawn. If plots are too small, there is competition between them. In 1924, GOSSET recommended the outer rows of all plots be discarded at harvest. Systematic layouts ABC . . . ABC . . . led to bias in estimates of treatment differences. He recommended the half-drill strip.

In 1938, he published a very important paper. Until this point he and FISHER had corresponded and agreed. He clearly showed that "balance" for a suspected trend leads to smaller bias in estimates of treatment effects but tends to overestimate error. This agrees with much of FISHER, but GOSSET concluded that one should balance, while FISHER said that one should block and randomize. GOSSET used uniformity data. He calculated the variance ratio for a particular layout and some particular values of the variance of treatment effects (assuming an additive model). He was assessing power, whereas WELCH and PITMAN had been doing randomization tests on uniformity data, so they were assessing significance.

After 1920, the biometrical genetics and statistics in breeding rapidly developed fast, so that a detailed discussion would burst the framework of the book.

E. S. PEARSON proposed the analysis of variance in cases of non-normal variation. In 1937, he assumed that the purpose of randomization is to do a randomization test. He pointed out that this test depends on the statistic used following some examples of FISHER. The so-called "z-test" was introduced for experiments with randomized blocks and Latin squares by WELCH (1937). He assumed arbitrary fixed plot effects additive with treatment effects. For randomized blocks, expectations of differences and of mean squares agree with those of normal theory. Because the plot effects are "fixed," the usual mean squares are not independent. The tail probabilities of the variance ratio may not be those given by normal theory, but several sets of uniformity data show good agreement. Actually he used as statistic the treatment sum of squares, calculated formulae for its mean and variance, and applied a normal approximation to that. For Latin squares, he randomized by choosing at random from among all Latin squares of the same size. The variance of the treatments sum of squares depends on the number of intercalates in the squares.

Later >>> F. YATES (1939) discussed the point that the purposes of randomization are to avoid accidental bias, to make the results credible, and to obtain unbiased estimators both of treatment differences and of their variances. Therefore, he was interested in estimation rather than testing. He supports GOSSET's criticisms of FISHER, but goes on to consider the possibility of randomizing designs, which have complicated ways of allowing for trend, such as the Latin square with balanced corners and the Latin square with split-plots.

With the introduction of complex multifactorial trials, additional problems of estimation were revealed. R. L. PLACKETT and J. P. BURMAN (1946) optimized symmetrical factorial experiments with all treatment factors equireplicate. They introduced what is now called "orthogonal arrays of strength two." For factors at two levels they used what are now called "Hadamard matrices," constructed by PAYLEY's method. For p levels with p prime, affine geometries via sets of mutually orthogonal Latin squares have to be used. N. L. JOHNSON (1948) proposed alternative systems in the analysis of variance and discussed how the randomization procedure can justify the assumption of a simple model, in a simple case.

In 1951, D. R. COX claimed more systematic experimental designs. Plots should be in a line, with a low-order polynomial trend. He wanted treatment contrasts to be orthogonal to this trend as far as possible. If the design is symmetric about its midpoint, then treatment contrasts are orthogonal to odd-order orthogonal polynomials and so trial-and-error solutions orthogonal to the quadratic orthogonal polynomial are orthogonal to cubic trend. H. D. PATTERSON (1952) favored the construction of balanced designs for experiments involving sequences of treatments, i.e., change-over designs for first-order residual effects. Each treatment should occur equally often in each period. Subjects form a balanced (complete or incomplete) block design. Each treatment follows each other treatment equally often.

Statistical methods for design and analysis have made a major contribution to improving experimental efficiency since FISHER and YATES established the basic principles of the experimental method in the 1920s and 1930s. These principles—to avoid bias, control haphazard error, take due account of treatment structure and experimental constraints, and ensure objectives are met—need to be interpreted for each situation. Experiments that appear to be similar may have very different objectives and thus require different approaches to design and analysis.

Despite the subject's long history, there are still many exciting opportunities for statistical research, as new fields of application open up and the objectives of experimental studies become increasingly complex. Recent advances in computer power have also led to a substantial increase in the range of feasible analytical techniques.

Much research has been stimulated by close links with plant-breeding and variety-testing organizations. This has ensured that the new techniques meet user needs and that they are rapidly taken up for routine use. The book *Statistical Methods for Plant Variety Evaluation* by R. KEMPTON and P. FOX (1997) has heavily contributed to the development and application of methods in sensory analysis. Computer algorithms, such as "GenStat," partially developed by researchers of traditional Rothamsted Experimental Station (United Kingdom) for generating and analyzing designs, have been incorporated into statistical packages and are widely utilized in statistical analysis of field and genetic experiments.

At present, plant breeding involves assessing the relative performance of large numbers of genotypes over several stages of selection. Trials may cover large areas of land, and comparisons between genotypes are then likely to be distorted by fertility trends or other sources of plot-to-plot variation. Much work has been done on modeling the variation in plot response and investigating control methods, which will lead to increased precision of genotype comparisons. The use of small blocks of plots, with genotypes allocated to blocks in an alpha design, was proposed 1990s. The latest developments include blocking in two dimensions using row and column designs, which we have found provide substantial additional gains in precision. Alpha designs are now widely used by national and international organizations throughout the world. Computer programs for generating alpha designs and analyzing their results are available from different providers.

Spatial modeling has been enthusiastically promoted since the early 1980s as an alternative to block designs for controlling plot variation. Current work is focused on the use of generalized additive models, which use the data directly to represent spatial variation across the trial, without assuming that these trends follow any fixed functional form.

Interference occurs when the response to a treatment applied to an experimental unit is affected by treatments on neighboring units. In variety testing, for example, interplot competition can lead to a reduction in the yield of shorter varieties due to shading from taller varieties in adjacent plots. This can produce substantial biases in the prediction of relative performance of varieties when in commercial production. Another application relates to the visual scoring of samples, for example, when screening plants for disease.

Many other proposals and tests were continuously introduced. They cannot be ruled out here. However, they were always associated with progress in breeding and complexity/specificity of field and laboratory testing.

3.10.3. Bioinformatics

A majority of economically important plant traits, such as grain or forage yield, can be classified as multigenic or quantitative. Even traits considered to be more simply inherited, such as disease resistance, may be "semiquantitative," for which trait expression is governed by several genes (e.g., a major gene plus several modifiers). The challenge to strategically use new technology (such as DNA-based markers) to increase the contribution of "science" to the "art plus science" equation for plant improvement therefore applies to most, if not all, traits of importance in plant breeding programs.

Historically, early researchers in quantitative genetics questioned whether the inheritance of these continuously distributed traits was Mendelian (COMSTOCK, 1978). The answer to this question has major implications in the consideration of the use of markers for plant breeding programs. During the past century, plant geneticists have compiled convincing evidence that Mendelian principles apply to quantitative as well as to qualitative traits. This evidence has also shaped the general model that embraces the multiple-factor hypothesis for quantitative traits, i.e., with genes located in chromosomes and hence sometimes linked, and incomplete heritability because of the contribution of environmental factors to total phenotypic variation.

The impact of the information revolution in crop improvement can be partially assessed by counting the number of publications indexed in *Plant Breeding Abstracts* (CAB International, Wallingford, U.K.). There was about a thirty-one-fold increase in publications between 1930 and 2005. In the 1970s indexed publications in plant breeding exceeded 10,000 per year. More publications and easy means for retrieving this information accounted for such growth of knowledge dissemination in plant genetics and breeding. Today, rapid information exchange has been facilitated by electronic mail and Internet access to electronic publications. Information technology and DNA science are beginning to fuse into a single operation. Computers decipher and organize the vast amount of genetic information that may become "the raw resource of the emerging biotech economy" (RIFKIN, 1998). Scientists working in the new field of "bioinformatics" are developing biological data banks to download the genetic information accumulated during millions of years of life evolution and perhaps help to reconstruct some of the living organisms of the natural world (ORTIZ, 1998).

Bioinformatics may be simply described as the data repositories, data mining, and analysis tools designed to interpret the genome data that is currently available, along with the data that will soon deluge the plant science community. Current public repositories include sequence databases, species-specific genome databases, and the germ-plasm databases. The sequence databases, such as "GenBank" (nucleotides) and "SwissProt" (amino acids), are robust resources for sequence deposition and sequence analyses. They provide powerful online tools for sequence analysis and searching, where searches can be made for motifs and secondary structure as well as for amino acid or nucleotide similarities. Sequence databases can support record-to-record links to the species-specific databases. However, these links need to be specified by curators of the species-specific databases.

Species-specific databases exist for major crops, such as soybean, maize, wheat, rye, barley, oat, and rice and for model organisms such as *Arabidopsis*. These genome databases integrate the map data for the species and provide documentation on the functionality of the genome. The data include the physical and genetic maps, clones and primers, QTLs, trait variances, references, images of pest and stress responses, and mutant phenotypes. They access the sequence information curated in the central sequence databases. The species-specific databases require scientific curation to ensure data quality, uniformity of gene and allele nomenclature, and accurate integration of data. The germplasm databases catalog information on available seed resources, along with certain agronomic and quality trait data. They do not presently contain any genome information, but linking to the genome databases is under investigation.

Chapter 4

Biotechnology, Genetic Engineering, and Plant Improvement

Advances in biology augur a third agricultural revolution involving biotechnology, a catchall term that includes both cell and DNA manipulation. D. N. DUVICK designated the hybrid maize production of the 1930s as biotechnology. A conventional baseline for the biotechnological revolution is 1953, the year of the >>> J. WATSON and >>> F. CRICK paper on the structure of DNA. However, the biotechnological revolution has no precise beginning because science is cumulative. One pathway developed from a series of investigations into gene function and structure. Another developed from the culture and physiology of cells using microbial techniques. It was stressed several times in the previous chapters that recent advances in biotechnology often have their roots in ancient epochs.

The beginnings of inquiries into gene structure can be traced to the 1860s, when Swiss biologist J. F. MIESCHER (1844-1895) described a substance he called "nuclein" that was derived from pus scraped from surgical bandages and later found in fish sperm. In 1889, R. ALTMANN (1852-1900), a student of MIESCHER, split nuclein into protein and a substance he named "nucleic acid." Two distinct kinds of nucleic acids were found in thymus and yeast. ASCOLI, in 1900, and LEVENE (1869-1940), in 1903, demonstrated the presence of adenine, cytosine, guanine, and thymine in thymus nucleic acid (now known as DNA) and, with uracil replacing thymine, in yeast nucleic acid (RNA). The original assumption that those bases were present in equal amounts and thus formed a "stupid" molecule proved faulty, but E. CHARGAFF (1905-2002), in 1950, demon-

Concise Encyclopedia of Crop Improvement
© 2007 by The Haworth Press, Inc. All rights reserved.
doi:10.1300/5891_04

strated a key equality. In molar amount, cytosine equaled guanine, and adenine equaled thymine. This suggestion was basic to the complementary replication of DNA and to the discovery of the structure of the double helix.

Genetic studies of the biochemistry of gene products date to 1902, when the British physician A. GARROD (1857-1936) demonstrated that the human disease alkaptonuria was inherited and due to an alteration in nitrogen metabolism. In a paper entitled "Inborn Errors of Metabolism," alkaptonuria was established as a gene-induced enzymatic block. This prescient study affected the course of biochemistry but remained unappreciated, if not unread, by early geneticists. The genetic investigations of metabolism from 1900 to 1950 formed a subdiscipline: biochemical genetics. The one-gene-one-enzyme model predicted by GARROD was established as dogma in the new catechism of biochemical genetics by >>> G. W. BEADLE and E. L. TATUM. The utilization of bacteria and bacteriophages, with new and powerful techniques for recombinational analysis, changed the concept of the particulate gene. Long considered to be analogous to a bead on a string, the gene was finally shown to be more of a long molecule, as first proposed by R. GOLDSCHMIDT (1878-1958).

As a genetic subject, the emergence of the power of microbial systems and the rise of bacteriophage, with its twenty-minute generation time, altered the experimental approaches. A clear distinction arose between the old and the new genetics (>>> M. DELBRÜCK, F. CRICK, S. BENZER, J. LEDERBERG, J. WATSON, and M. NIREMBERG). Analysis of the transformation principle in *Pneumococcus* by O. AVERY, C. McLEOD, and M. McCARTY, and, in 1952, the subsequent phage manipulations by A. D. HERSHEY (1908-1997) and M. CHASE (1930-2003), proved that the genetic material was DNA. However, details on its ability to replicate and to affect protein synthesis were unknown until the famous paper by WATSON and CRICK (1953). The resolution of the structure of DNA was followed by a race that unraveled the genetic code and by the discovery of restriction endonucleases that snip gene sequences and plasmid vectors that transfer them across barriers considered unbridgeable by even the most credulous imaginations. In 1973, it became feasible to isolate genes, and from 1980 onward, the application of DNA recombinant techniques stimulated a boom of research.

By conventional breeding, developing a new variety can take up to fifteen years for wheat, eighteen years for potatoes, and even longer for some crops. Because of the tremendous impact of biotechnology, the Nobel Peace Laureate N. BORLAUG (1997) suggested the application of new biotechniques, in addition to conventional plant breeding, to boost yields of common crops that feed the world. Therefore, the scope of conventional plant breeding has increased with improvements in technology. In the laboratory, chemical and mechanical techniques are used to speed up the selection process and remove natural barriers to cross-fertilization.

4.1. IN VITRO TECHNIQUES

Tissue culture for breeding application was developed in the 1950s, and it became popular in the 1960s. Today, micropropagation and *in vitro* conservation are standard techniques in most important crops, especially those with vegetative propagation. The early history of cloning as a technical object began in two different experimental systems during the first two decades of the 20th century. Like the concept of the gene, the concept of the clone was first introduced to biology at the beginning of the 20th century. Whereas the gene concept referred to an abstract or even an ideal unit (according to W. JOHANNSEN's use of the term), the concept of the clone referred from the beginning to a concrete material object. In 1905, H. J. WEBBER, a botanist from the Plant Breeding Laboratory of the USDA introduced the term "clone" to designate groups of plants that are propagated by the use of any form of vegetative parts and which are simply parts of the same individual seedling (see HODGSON, 2002). Though the concept was first developed in the context of horticultural breeding, the clone soon became something that could be called a technical object of different experimental systems. In a broad sense, in vitro techniques include the following:

1. Seed culture
2. Embryo culture
3. Ovary or ovule culture
4. Anther or microspore culture (cf. Photo 3.4)
5. In vitro pollination
6. In vitro fertilization

7. Organ culture
8. Shoot apical meristem culture
9. Somatic embryogenesis, organogenesis, and enhanced axillary budding
10. Callus culture
11. In vitro production of secondary metabolites
12. Cell culture and in vitro selection at cellular level
13. Genetic and epigenetic somaclonal variations
14. In vitro mutagenesis
15. Protoplast isolation, culture, and fusion
16. Genetic transformation
17. In vitro flowering
18. Micrografting
19. Cryopreservation or storage at low temperature
20. Culture of protoplasts, cells, tissues, and organs
21. Culture of hairy roots

The history of plant tissue culture begins with the concept of the cell theory given independently by M. J. SCHLEIDEN (1838) and T. SCHWANN (1839), which implied that the cell is a functional unit. Pioneering studies of in vitro culture of monocot plant organs and tissues by G. HABERLANDT in 1902 predicted that the notion of producing plants from cultured cells would provide final confirmation of the cell theory (i.e., the cellular totipotency). In 1958, F. C. STEWARD, M. O. MAPES, and K. MEARS demonstrated it. The team was the first to be able to transform a carrot cell line in some "artificial embryos," later called "somatic embryos."

In the past fifteen years, REDENBAUGH (1993) has developed very interesting possibility of the somatic embryogenesis technology. Since 1958 when the first plant embryos were obtained from somatic tissues of carrots cultured in vitro, ever-increasing numbers of species and tissues have been induced to form somatic embryos. REDENBAUGH was able to encapsulate somatic embryos within hydrogel coatings (sodium alginate), producing single-embryo artificial seeds. To date, some improvements offer the possibility to directly plant the artificial seeds in the greenhouse on special substrates. The technique of using artificial seeds was successfully applied in alfalfa, asparagus, bamboo, caraway, carrot, celery, cinnamon, coffee, eggplant, geranium, ginger, horseradish, lettuce, lily, papaya, rice, sandalwood, tangerine, spruce, olive, and vanilla by coating somatic embryos with alginate, polyox, polyethyleneglycol, or cellulose. There was slow but continuous progress. However, in the future the methodology will provide a good technique to reduce the cost of transplants.

In 1922, W. J. ROBBINS introduced procedures for the culture of roots and L. KNUDSON developed the aseptic germination of the embryo-like seed of orchids. In 1934, P. R. WHITE (1934) reported for the first time successful continuous cultures of tomato root tip in liquid culture, and he obtained indefinite growth. In 1939, he was able to regenerate the first shoot from tobacco callus. During the same year, NOBECOURT cultivated plant tissues of *Nicotiana tabacum* and *Salpiglossis sinuata* for unlimited periods.

The first plantlet formation in vitro was reported as early as the 1940s (e.g., in *Tropaeolum* and *Lupinus*). CAPLIN and STEWARD (1948) used coconut milk for the first time in cultures of single cells and introduced various types of vessels for culture work, such as rotating nipple flasks. R. J. GAUTHERET (1955) established habituated cultures of tumor cells and discussed the importance of light and temperature for root growth. The French botanists G. MOREL and R. A. WETMORE (1951) were the first to culture monocotyledonous tissue. They used meristem tip culture for elimination of virus disease of orchids and discovered two unique opines of crown gall tissues. In orchids, plantlet formation was reported by MOREL during the 1960s, which became a commercially viable program.

The breakthrough in plant cell and tissue culture arose from a series of physiological investigations, principally by F. SKOOG and his co-workers. They examined growth-regulating substances, including vitamins, hormones (particularly auxin and cytokinin), and organic complexes such as liquid coconut endosperm. Break throughs also came with the development of generalized tissue culture media by P. R. WHITE in the 1930s and 1940s and most successfully by T. MURASHIGE and SKOOG (1962).

The demonstration of asexual embryos initiated in the cultures of carrot root cells in 1958 by J. REINERT, F. C. STEWARD, and K. MEARS was a confirmation of the concept of cell totipotency (i.e., that each living cell contains all the genetic information). REINERT (1959) showed bipolar embryo formation in carrot and embryogenesis, the possibility of cryopreservation of cells and full regeneration of plantlets. VASIL and HILDEBRANDT (1965) published a landmark work on single isolated cells of tobacco, which paved the way for cell cultures in genetics. It has contributed significantly to cell and protoplast cultures as well as somatic embryogenesis of cereals.

Extensive investigations continue to explore the potential of cell and tissue culture as an adjunct to crop improvement. Techniques include embryo rescue, freeing plants from viruses and other pathogens, haploid induction, cryogenic storage of cells, meristem for germplasm preservation, the creation of new nuclear and cytoplasmic hybrids via protoplast fusion, and the exploitation of changes, or somaclonal variation. It was recognized that cell and tissue culture technology would be required as an intermediary for recombinant DNA technology.

4.1.1. Embryo Rescue

Failure to produce a hybrid may be due to pre- or postfertilization incompatibility. If fertilization is possible between two species or genera, the hybrid embryo aborts before maturation. When the cross is incompatible after fertilization, the embryo resulting from an interspecific or intergeneric cross can be rescued and cultured to produce a whole plant.

In 1873, P. von TIEGHEIM tried to culture immature embryos of *Mirabilis jalapa* on potato starch. In 1904, E. HANNIG used several crucifers to make another attempt in embryo culture. Seventeen years later, M. MOLLIARD cultivated fragments of different plant embryos, while L. KNUDSON was able to demonstrate asymbiotic germination of orchid seeds in 1922. Crop species were first embryo-cultured in 1925 by F. LAIBACH (F1 embryos of interspecific crosses in *Linum* species). C. R. LARUE (1936) published a report on embryo culture of gymnosperms. The usefulness of coconut milk for growth and development of very young *Datura* embryos was discovered by J. OVERBEEK in 1941. By 1951, excised ovaries were cultured in vitro (NITSCH, 1951), and K. KANTA demonstrated in 1960 the first test tube fertilization in *Papaver rhoeas*.

In 1970s, this technique was used to produce new rice for Africa: an interspecific cross of Asian rice *(Oryza sativa)* and African rice *(O. glaberrima)*. However, its application is mainly restricted to pre-breeding programs, where exotic germplasm has to be introgressed to advanced breeding strains and progamous or postgamous incompatibility has to be managed. An exception was demonstrated by >>> K. J. KASHA and K. N. KAO (1970). They applied the embryo rescue method in *Hordeum vulgare* × *H. bulbosum* crosses, although KONZAK et al. (1951) had already used the technique for the same combination. By chance, they discovered spontaneous chromosome elimination leading to haploid seedlings. This technique subsequently became routine in the haploid production of barley breeding programs (see Chapter 3.8).

4.1.2. Cell Fusion and Somatic Hybridization

As in sexual hybridization, cell fusion techniques contribute to the combination of whole genomes. When sexual hybridization fails, somatic hybridization can be an alternative. Different types of cells are dissociated from tissues, walls are stripped from the cells, and the resulting protoplasts are in vitro cultured and regenerated into intact plants. This technique can produce novel combinations of nuclei, mitochondria, or chloroplasts.

Protoplasts are the smallest units capable of regenerating a whole plant. Therefore, protoplasts cultures can serve to enlarge genetic variability by introducing somaclonal variation (see Chapter 3.5.2). However, the main characteristic of protoplasts is their capacity to fuse and to produce hybrids or cybrids (i.e., new organisms most often unknown in nature). Although in nature, cybrids were already made millions of years ago by spontaneous cell fusion of algae and cyanobacteria resulting in plants able to perform photosynthesis. In 1909, the first man-made fusion of plant protoplasts was attempted by E. KÜSTER, but the products failed to survive. However, P. S. CARLSON et al. (1972) succeeded. Protoplasts of *Nicotiana glauca* and *N. langsdorffii* were isolated, fused, and induced to regenerate into plants. The somatic hybrids were recovered from a mixed population of parental and fused protoplasts by a selective screening method that relies on differential growth of the hybrid on defined culture media. The biochemical and morphological characteristics of the somatically produced hybrid were identical to those of the sexually produced amphiploid.

The first successful electrical fusion was demonstrated by the German U. ZIMMERMANN (1982) with the use of mesophyll protoplasts of *Kalanchoe* or *Avena* species. In 1983, G. PELLETIER and co-workers produced an intergeneric cybrid using radish and rape. Indeed, protoplasts allow transgressing of the barrier of the botanical species or genus.

Naked protoplasts can also accept without rejection external elements containing genetic information. The first protoplast fusion application for breeding was a cytoplasmic transfer from one genotype to another to induce male sterility (cytoplasmic male sterility, or CMS) from mitochondrial origin. These male sterile hybrids are interesting to produce F1 hybrids (e.g., in *Brassiceae*). Another cytoplasmic transfer, from *Solanum nigrum* to tomato, concerns the introduction of chloroplastic DNA to induce herbicide (atrazine) resistance (JAIN, SHAHIN, and SUN, 1988). By asymmetric cell fusion techniques, the first commercial tobacco male sterile line was developed. Such asymmetric male sterile lines for commercial exploitation of F1 hybrids have also been bred in carrots, cabbages, and eggplants (NAKAJIMA, 1991).

In 1989, in Sweden >>> SJÖDIN and GLIMELIUS successfully transferred resistance against *Phoma lingam* to *Brassica napus* by asymmetric somatic hybridization combined with toxin selection. Japan and China are particularly involved in this work. During the past twenty years, successful protoplast cultures were reported in more than 100 plant species. Plants regenerated from protoplasts have been obtained in about 60 species, including vegetables, medicinal plants, legumes, and other economic crops, as well as tree species such as poplar, elm, and rubber trees (Exhibit 4.1). Using cell fusion techniques, novel citrus hybrid varieties, such as "Oretachi" (orange plus trifoliate orange), "Shuvel" (Satsuma mandarin plus navel orange), "Gravel" (grapefruit plus navel orange), "Murrel" (murcott plus navel orange), and "Yuvel" (Yuzu plus navel orange) were developed. However, in several cases, the regeneration frequency is low and has to be improved.

EXHIBIT 4.1. Some examples of regenerated plants derived from induced somatic protoplast fusion of interspecific and intergeneric combinations.

Food crops

Hordeum vulgare + *Daucus carota; Triticum aestivum* + *Agropyron elongatum; T. aestivum* + *Haynaldia villosa; T.* + *Leymus chinensis; Oryza sativa* + *Echinochloa oryzicola; O. sativa* + *O. brachyantha; O. sativa* + *O. eichingeri; O. sativa* + *O. officinalis; O. sativa* + *O. perrieri; O. sativa* + *Hordeum vulgare; O. sativa* + *Porteresia coarctata; Solanum tuberosum* + *S. phureja; S. tuberosum* + *S. papita; S. tuberosum* + *S. Bulbocastanum*

Fodder plants

Medicago sativa + *M. arborea; M. sativa* + *M. coerulea; M. sativa* + *M. falcata; M. sativa* + *Lotus corniculatus*

Vegetables

Brassica campestris + *B. oleracea; B. oleracea* + *Raphanus sativa; B. oleracea* + *Moricandia arvensis; B. oleracea* + *Sinapis turgida; B. napus* + *B. nigra; B. oxyrrhina* + *B. campestris; B. napus* + *Raphanus sativa; Daucus carota* + *D. cupillifolius; Diplotaxis catholica* + *B. juncea; Erucastrum gallicum* + *B. campestris; Lactuca sativa* + *L. virosa; Lycopersicon esculentum* + *Solanum muricatum, S. ochranthum; L. esculentum* + *S. melongena; Moricandia arvensis* + *B. juncea; Sinapis alba* + *B. juncea; S. commersonii, S. phureja* + *S. tuberosum; S. melangena* + *S. sanitwongsei; S. melangena* + *S. integrifolium; Trachystoma balli* + *B. juncea*

Ornamentals plants

Petunia hybrida + *P. parodii*

Fruit and forest trees

Citrus autrantium + *C. reticulata; C. sinensis* + *C. paradisi; C. sinensis* + *C. unshiu; C. sinensis* + *C. sinensis x tangerina; C. sinensis* + *C. sinensis x Poncirus trifoliata; C. sinensis* + *Poncirus trifoliata*

Industrial crops

Nicotiana tabacum + *N. africana; N. tabacum* + *N. debney; N. tabacum* + *N. rependa; N. sylvestris* + *N. plumbaginifolia; N. tabacum* + *Atropa belladonna*

Source: Adapted from NAKAJIMA, 1991, and NAGATA and BAJAJ, 2001.

4.1.3. Virus Freeing

Twenty years after HABERLANDT's fruitless in vitro studies, his successors developed techniques useful in investigations of plant development (see Chapter 4.1). The character of studies dictated the use of small tissue fragments as experimental objects to have full control over their living processes. The small size of *explants* forced the introduction of culture sterility to protect them from invasions of microorganisms inhabiting plant surfaces and those present in the air. The first real successes in keeping small fragments of plant tissues alive in vitro for a prolonged time came with the discovery of auxins in the late 1930s. The addition of indoleacetic acid (IAA) to tissue culture media enabled cultivation of callus, roots, and later shoots for long periods of time.

In the 1950s, new plant hormones such as cytokinins and gibberellins were discovered. Now, researchers were in almost full control of tissue growth and development in sterile environments. Knowledge of plant development reached the level that permitted commercial applications of tissue culture. It started with freeing plants from viruses. This could be achieved by isolation of very small shoot apices (meristems) and growing them on sterile media, which provided for their transformation into small plants. These plants could then be taken out of sterile culture vessels and established in standard (unsterile) horticultural environments.

In 1952, MOREL and MARTIN were successful in regenerating a virus-free dahlia plant by the excision of some meristematic domes from virus-infected shoots. Three years later, the same authors were able to eliminate viruses A and Y from a virus-infected potato. SEMAL and LEPOIVRE (1992) reported that a virus-free sweet potato was producing 40 t/ha in China in comparison with the 20 t/ha produced before meristem culture. Virus eradication depends on several parameters. However, to take advantage of the nonuniform and imperfect virus distribution in the host plant body, the size of the excised meristem should be as small as possible. Only tips between 0.2 and 0.5 mm most frequently produce virus-free carnation plants. The explants smaller than 0.2 mm cannot survive, and those larger than 0.7 mm produce plants that still contain mottled viruses.

Today, no definitive explanation can be given to understand this virus eradication. Various explanations have been given, such as an absence of plasmodesm in the meristematic domes; competition between synthesis of nucleoproteins for cellular division and viral replication; inhibitor substances; absence of enzymes present only in the cells of the meristematic zones; and suppression by excision of small meristematic domes. This last proposal could explain why some potato plants showing virus particles in the meristematic domes could regenerate a virus-free plant.

Dahlias, carnations, chrysanthemums, and cymbidiums were the first plants freed from viruses through meristem culture. The development of cymbidium in tissue culture was more complicated. Small shoot tips, when placed on media, produced globular structures called protocorms. Protocorms could easily be stimulated to produce new protocorms by mechanical separation. In this way, the stock of protocorms grown in vitro could easily be increased from one to a few million during one year of culture.

Producers of planting material expected big money in the application of this propagation method, and this stimulated fast development of commercial in vitro propagation methods in the late 1960s and early 1970s. To lower costs of this labor-intensive production, specialized equipment and instruments were developed. Laminar airflow cabinets, providing bacteria-free and fungi-free work environments, were introduced to tissue culture production in the early 1970s. At the same time, researchers coined the term "micropropagation" denoting vegetative propagation in vitro. Nowadays, hundreds of thousands of hectares are planted with virus-free plantlets.

4.1.4. Micropropagation

Micropropagation is another technique of great significance to agriculture, which is based on in vitro culture of plant cells. Each cell of a plant has the potency to be regenerated into a complete plant by tissue culture techniques. This cell property is known as totipotency. There are two basic strategies used to multiply plants in vitro:

1. Proliferation of preformed buds
2. Bud induction, termed adventitious regeneration

The micropropagation process is conveniently divided into five stages: (1) stock plant selection, (2) establishment of aseptic cultures, (3) multiplication, (4) preparation for microplant establishment, and (5) microplant establishment.

The 1970s were a decade of boom in horticultural micropropagation. It was facilitated not only by technical improvements but also by the work of many researchers. The most important discoveries, from a micropropagation point of view, were investigations into the role of growth regulators in shoot apical dominance and shoot branching. Stimulation of shoot branching is the core of the most reliable micropropagation methods of today.

From among hundreds of researchers active in the field at that time, T. MURASHIGE and his students from the University of California distinguished themselves by developing practical and theoretical foundations for today's micropropagation. They outlined the method, dividing it into four stages. They also developed commercial micropropagation methods of many important ornamental plants. Media developed by them in the early 1970s are still the most commonly used in many commercial laboratories.

The application of labor-intensive, high-tech, and, therefore, expensive micropropagation has been economically justified chiefly in the case of ornamental plants. This situation has not changed in the past twenty years. The need for sterility of cultures dictates the use of specialized and expensive equipment. The small dimensions of objects, and the need for their careful handling during subcultures on fresh media, lower labor efficiency. The cost of labor constitutes 60 to 70 percent of total micropropagation costs.

Reduction of micropropagation costs can be achieved by its mechanization and automation. It is a difficult task because the structure of plant objects grown in vitro is very complicated. Only a few plants, like eucalypti or potatoes, produce simple elongated shoots with clearly visible internodes and leaves. Most ornamental plants branch intensively during culture, producing thickets of entangled shoots of different sizes. Efficient automatic division of such structures is impossible with present levels of knowledge in three-dimensional computerized vision systems. In addition, difficulties in sterilizing complicated robots and accessory equipment must be added. Using this technique, hundreds and thousands of seedlings can be generated in a very short time, starting from a limited number of explants. This method of multiplication has two major advantages:

1. First, it accelerates multiplication of new elite cultivars. For example, after its release, a new sugarcane cultivar may take five to ten years before sufficient planting material may be generated for wide coverage in a farmer's field. However, using micropropagation techniques a comparable level of multiplication may be achieved within two years.
2. Second, one can get disease free planting material because of the use of aseptic conditions during tissue culture (see Chapter 4.1.3).

There are also reports of rejuvenation and increased vigor following tissue culture. This can help in the rescue of important varieties which are on the verge of extinction due to disease attack. In species such as daffodils and gladioli, micropropagation techniques are being used to speed up the release of new varieties. In strawberries, millions of plants can be produced from a single mother plant in one year.

Another application is in vitro selection. In maize, callus cultures were established resistant to T-toxin of *Helminthosporium maydis* (GENGENBACH and GREEN, 1975). Similar experiments were carried out with potatoes against *Phytophthora infestans* (1979), with barley (1987) alfalfa against *Fusarium oxysporum* f. sp. *medicagnis* (1984), in *Brassica napus* against *Phoma lingam* (1982), and in tobacco against *Pseudomonas* and *Alternaria* toxins (1983).

Somatic embryogenesis is also being commercially applied. Synthetic seeds ("synseed") of rice and vegetables have been developed by the private sector in Japan. Triploid clones of rubber produced through somatic embryogenesis have outyielded diploid standard clones by about 20 percent in China. To facilitate reforestation and promote social forestry and agroforestry, India has established two tissue culture pilot plants, one at the National Chemical Laboratory, Pune, and the other at the Tata Energy Research Institute, New Delhi, with production capacities of five to ten million propagules and/or seedlings of elite and/or plus trees of several important species, such as *Eucalyptus tereticornis, E. camadulensis, Tectona grandis, Dendrocolamus strictus, Populus deltoides, Bambusa vulgaris,* and *B. tulda.* Micropropagation of teak, rattan, eucalyptus, bamboo, and other tree species has been adopted in several other countries.

Novel breeding strategies are also being tried in potato breeding by using microtubers for propagation and cropping, despite strategies based on genotypic selection. It concerns the so-called "true potato seed" (TPS). However, the breeders are still skeptical about the place of TPS in highly developed markets of Europe and North America, although there are efforts to develop sexually propagated varieties. The basis for this scheme rests on the generation 2n gametes by first meiotic division restitution (FDR).

4.1.5 In Vitro Conservation of Germplasm

Several plant species, including a few commercial species, produce recalcitrant seeds; hence it is difficult to conserve them through seeds. Furthermore, some species are shy seed bearers and even fail to produce seeds. In addition, living collections of clonally propagated perennial crops face the problems of maintenance of heterozygous and heterogeneous populations, the long life cycle and large space required, and a high possibility of exposure to the threats of pests and diseases and other biotic and abiotic stresses. To circumvent these difficulties, in vitro conservation of vegetatively propagated and recalcitrant seed-producing species is being increasingly adopted as a complement to other methods of conservation. For short- to medium-term storage (working and active collections), the slow growth method is used, whereas for long-term storage (base collections), cryopreservation is the method adopted. In 1976, M. SEIBERT showed the first success of shoot induction from cryopreserved shoot tips of carnation.

The sequencing of crop genomes opened new frontiers in conservation of plant biodiversity and its genetic enhancement. The advances in gene isolation and sequencing in many plant species allows to envisage that, within a few years, gene bank curators may replace their large cold stores of seeds with crop DNA sequences that will be electronically stored. The characterization of plant genomes will ultimately create a true gene bank that should possess a large and accessible gene inventory of today's noncharacterized crop gene pools. Of course, seed banks of comprehensively investigated stocks should remain because geneticists and plant breeders, the main users of gene banks, will need this germplasm for their work. Genomics may also accelerate the utilization of candidate genes available at these gene banks through transformation without barriers across plant species or other living kingdoms.

4.2. MOLECULAR TECHNIQUES IN PLANT BREEDING

At the beginning of the 1980s, genetic engineering of plants remained a promise of the future, although gene transfer had already been achieved earlier in a bacterium. The first transgenic plant, a tobacco accession resistant to an antibiotic, was reported in 1983. Transgenic crops with herbicide, virus or insect resistance, delayed fruit ripening, male sterility, and new chemical composition have been released to the market with growing intensity. For example, the *antisense* strategy was applied by CALGENE Inc. in 1994, when the first commercial transgenic plant was created, a long shelf-life tomato, by the suppression of polygalacturonase activity due to an antisense gene (SMITH et al., 1988). However, this "Flavr Savr" tomato variety was removed from the trade three years later because of its disease susceptibility and its lack of productivity. Later, other tomato varieties with long storage qualities were obtained by the utilization of an *antisense RNA* inhibition of ACC synthase or ACC oxidase, two ethylene precursors.

The *antisense technique* was also used to reduce the lignification of woody plants, by blocking the enzymes involved in the precursor of lignin biosynthesis. Another interesting application was the induction of white flowers in petunia and in different other ornamental plants by the suppression of chalcone synthase activity. "Shutting off" genes coding for undesired characteristics will be an attractive application of transgenics in future crop improvement.

4.2.1. Marker-Assisted Selection

For tracing quantitative traits, morphological markers were first applied. One such approach was the development of mutations at the loci controlling plant morphology of maize (STADLER, 1929). The variations represented through these mutations were observed as altered plant phenotypes that ranged from pigment differences and gross changes in development to disease resistance response. Nevertheless, morphological markers have not been used extensively in practical breeding because of the limited availability, environmental variation, and more or less tight linkage to the target character (see Table 4.1).

Allozymes were available as the first biochemical markers in the 1960s. Population geneticists took advantage of such marker system for their early research. Allozymes generally differ due to the substitution of a single amino acid of different charge at a locus. Such changes in amino acid composition often alter the charge or, less often, the conformation of the enzyme. This leads to a change in electrophoretic mobility of the enzymes and thus provides an extremely

TABLE 4.1. Types and characteristics of different markers in breeding and genetics.

Character-istics	Morpho-logical traits	Isoenzymes	RFLPs	AFLPs RAPDs	SSRs (Microsatellites or Simple Sequence Repeats)
Number of loci	Limited	Limited	Almost unlimited	Unlimited	High
Inheritance	Dominant	Codominant	Codominant	Codominant or dominant	Codominant
Positive features	Visible	Easy to detect	Robust; reliable; transferable across populations	Quick assay with many markers; multiple loci; small amounts of DNA needed; highly polymorphic; inexpensive;	Well distributed within the genome, many polymorphisms
Negative features	Possibly negative linkage to other characters	Possibly tissue-specific	Radioactivity required; time-consuming; laborious; rather expensive; large amounts of DNA needed; limited polymorphism (especially in related lines)	High basic investment; large amounts of DNA needed; patented; complicated methodology; problems with reproducibility; generally not transferable	Long development of the marker; expensive

useful method of evaluating genetic differences among those that appeared prominent (see Table 4.1).

It was no later than 1974 that C. M. RICK and J. F. FOBES, and in 1980 H. P. MEDINA-FINHO, established a tight genetic linkage between a nematode resistance gene and an *Aps1* isozyme allele in tomato, which opened the avenue of tagging genes of agronomic importance. S. TANKSLEY and collaborators (1984) found peroxidases closely associated with male sterility or self-incompatibility in tomatoes. Soon after, similar markers were detected in rye, apple, maize, beans, and so on. The effect of isozymes and other proteins on a plant's phenotype is usually neutral, and both of them are often expressed codominantly, making the discrimination possible between homozygote and heterozygote. However, because of the limited number of protein and isozyme markers and because of the requirement of a different protocol for each isozyme system, their utilization was also very limited in plant breeding programs.

The discovery of enzymes, which cleave DNA at specific sequences and subsequently ligate to extrachromosomal DNAs of bacteria, permitted gene replication in a bacterial host, a process known as gene cloning. In the 1970s, such enzymes and their *restriction fragment length polymorphisms* (RFLP), displayed by SOUTHERN blotting, were added to the toolbox of the geneticists. This marker system was developed in early 1980 (BOTSTEIN et al. 1980). RFLP markers are codominant and available in unlimited number because only a small (1,000 nucleotide base pair) fragment is used for cloning from genomes that may contain a billion or more base pairs linearly arranged along the chromosomes (see Table 4.1).

The *Taq polymerase* was found in the 1980s, and the *polymerase chain reaction* (PCR) was developed shortly afterward. Since then, marker-aided analysis based on PCR has become routine in plant genetic research and marker systems have shown their potential in plant breeding (PATERSON, 1996). With this technology, a new generation of DNA markers, such as randomly amplified polymorphic DNA (RAPD), sequence characterized amplified regions (SCARs), sequence tagged sites (STS), single polymorphic amplification test (SPLAT), variable number of tandem repeats (VNTRs), amplified fragment length polymorphism (so-called *AFLPs;* KARP et al., 1997), DNA amplification fingerprinting (DFA), single-strand conformational polymorphism (SSCP), single-nucleotide polymorphism (SNPs), microsatellites or short tandem repeats (STRs), DNA micro arrays, and rDNA ITS (SCHENA et al., 1995) were introduced into the modern plant breeding systems. Some of these markers are relatively simple, easy to use, automatable, often codominant, near infinite in number and are comparatively faster to assay (see Table 4.1).

Furthermore, new single-nucleotide polymorphic markers based on high-density DNA arrays, a technique known as "gene chips" (CHEE et al., 1996), have been developed. With gene chips, DNA belonging to thousands of genes can be arranged in small matrices (or chips) and probed with labeled cDNA from a tissue of choice. DNA chip technology uses microscopic arrays (or microarrays) of molecules immobilized on solid surfaces for biochemical analysis (LEMIEUX et al., 1998). An electronic device connected to a computer may read this information, which will facilitate marker-assisted selection in crop breeding. There are five eras in genetic marker evolution:

1. Morphology and cytology in early genetics, until the late 1950s
2. Protein and allozyme electrophoresis in the "prerecombinant DNA time" (1960s-1970s)
3. RFLP and minisatellites in the "pre-PCR age" (1970-1985)
4. Random amplified polymorphic DNA, microsatellites, expressed sequence tags, sequence tagged sites, and amplified fragment length polymorphism in the "oligoscene period" (1986-1995)
5. Complete DNA sequences with known or unknown function as well as complete protein catalogs in the current computer robotic cyber genetics generation (1996-present)

By 2005, the DNA of a number of organisms had been completely mapped, including bacteriophage, bacteria, yeast, nematode, or the plants *Arabidopsis thaliana* and poplar. Recent projects concern major crop species, such as maize, rice, and wheat. Gene maps of individual chromosomes carrying important clusters of genes are available in tomato, maize, rice, and wheat. Analysis of gene function indicates that all living organisms have genes in common. Soon all our major crop plants will be mapped. The name of the next emerging field has already been coined: "proteonomics," which will unravel the protein changes involved with gene function and development (see the following text).

Although there have been numerous mapping and QTL mapping studies for a wide range of traits in diverse crop species, relatively few markers have actually been implemented in practical breeding programs. The main reason for this lack of adoption is that the markers used have not been reliable in predicting the desired phenotype. In many cases, this would be attributable to a low accuracy of mapping studies or inadequate validation. However, despite the lack of examples of marker-assisted selection being practiced, there is an optimism regarding the role in the future by leading researchers (COLLARD et al., 2005). The fields of application molecular markers are permanently growing. They are used in technical aspects of screenings, for characterization of germplasm, for cultivar improvement through DNA-content manipulation, for genetic dissection of traits, for *linkage map* construction, for detection of genetic linkages, for marker to gene (map-based) cloning, for marker-assisted selection (MAS), or for commercial variety identification, protection, and purification.

Frequent application of such marker systems transformed the conventional plant breeding programs into MAS breeding or plant molecular breeding. However, before initiating large-scale utilization of markers in plant-breeding programs, it is still necessary to have clear concepts of genes and genetic markers, as well as the characteristics and purposes so that the technique can be used effectively. Although recent advances in molecular genetics have promised to revolutionize agricultural practices, there are several reasons why molecular genetics can never replace traditional methods of agricultural improvement. They should be integrated to obtain the maximum improvement in the economic value of domesticated populations. Their analytical results, as well as the more recent computer simulations and the limited empirical results, however, are encouraging and support the use of DNA-based markers to achieve substantial increases in the efficiency of artificial selection.

4.2.1.1. Plant Genomics

This new term, defined by the development of biotechnology, refers to the investigations of whole genomes by integrating genetics with informatics and automated systems. Genomic research aims to elucidate the structure, function, and evolution of past and present genomes (LIU, 1997). Some of the most dynamic fields concerning agriculture are the sequencing of plant genomes, comparative mapping across species with genetic markers, and objective-assisted breeding after identifying candidate genes or chromosome regions for further manipulations. Because of genomics, the concept of gene pools has been enlarged to include transgenes and native exotic gene pools that are becoming available through comparative analysis of plant biological repertoires (LEE, 1998).

Nowadays, the finding of new genes that add value to agricultural products seems to be very important in agri-business. Unique gene databases are being assembled by the industry with the massive amount of data generated by genomics research. A new term "biosource" was coined recently to refer to a fast and effective licensed technology of pinpointing genes. With this method, a "benign" virus infects a plant with a specific gene that allows researchers to observe its phenotype directly. Biosource replaces the standard time-consuming approach of first mapping a gene to subsequently determine its exact function. Gene identification in DNA libraries, coupled with biosource technology and an enhanced ability to put genes into plants, will be routine for improving crops in the next decades.

Understanding the biological traits of one species may enhance the ability to achieve high productivity or better product quality in another organism. Today, DNA markers and gene sequencing provide quantitative means to determine the extent of genetic diversity and to establish objective phylogenetic relationships among organisms. "Gene chips" and transposon tagging will provide new dimensions for investigating gene expression. Molecular biologists study not only individual genes but also how circuits of interacting genes in different pathways control the spectrum of genetic diversity in any crop species. For example, more information will be available on why plant resistance genes are clustered together or on what candidate genes should be considered when manipulating QTL for crop improvement.

Genomics may provide a means for the elucidation of important functions that are essential for crop adaptability. Regions of the world should be mapped by combining data of geographical information systems, crop performance, and genome characterization in each environment. In this way, plant breeders can develop new cultivars with the appropriate genes that improve fitness of the promising selections. Fine-tuning plant responses to distinct environments may enhance crop productivity. Development of cultivars with a wide range of adaptation will allow farming in marginal lands. Likewise, research advances in gene regulation, especially those processes concerning plant development patterns, will help breeders to fit genotypes in specific environments.

Photoperiod insensitivity, flowering initiation, *vernalization,* cold acclimation, heat tolerance, and host response to parasites and predators are some of the characteristics in which advanced knowledge may be acquired by combining molecular biology, plant physiology and anatomy, crop protection, and genomics. Multidisciplinary cooperation among researchers will provide the required holistic approach to facilitate research progress in these subjects.

4.2.2. Transgenic Crop Plants

All crops have been genetically modified from their original wild state by domestication, selection, and controlled breeding over long periods. What is the major difference between "old" and "new" methods of crop improvement? It is the number of genes transferred to the offspring in each case. On average, plants contain approximately 80,000 genes, which can recombine during the process of sexual hybridization. The offspring may therefore inherit around 1,000 new genes because of this recombination. This is equivalent to a 0.0125 percent change in the genome. By contrast, only one or two new genes are transferred during plant transformation. As this represents a 0.0025 percent change in the genetic information of the plant, it is argued that plant transformation provides a more precise approach to crop improvement than sexual hybridization.

When it became clear that DNA could be inserted into the DNA of higher plants by various techniques, including the gene gun, a new era of research began. The most promising vector for dicotyledonous plants has been the tumor-inducing plasmid of *Agrobacterium tumefaciens,* a bacterium that normally incorporates its DNA in the host as part of the infection process. Even genetic engineering is not new!

The important prerequisites for successful integration of the *Ti* plasmid DNA from *Agrobacterium tumefaciens* in higher plants were published by CHILTON et al. (1977). In 1979, MARTON et al. developed the cocultivation procedure for the genetic transformation of plant protoplasts with *A. tumefaciens,* and the first naked DNA transformation of protoplasts was realized by KRENS et al. (1982).

Transgenic plants were first created in the early 1980s by four groups working independently at Washington University in St. Louis, Missouri, the Rijksuniversiteit in Ghent, Belgium, Monsanto Company in St. Louis, Missouri, and the University of Wisconsin, St. Paul.

On the same day in January 1983, the first three groups announced at a conference in Miami that they had inserted bacterial genes into plants. The fourth group announced at a conference in Los Angeles in April 1983, that they had inserted a plant gene from one species into another.

The Washington University group, headed by M.-D. CHILTON (BEVAN, FLAVELL, and CHILTON, 1983), had produced cells of *Nicotiana plumbaginifolia,* a close relative of ordinary tobacco, that were resistant to the antibiotic kanamycin (FRAMOND et al., 1983). J. SCHELL and M. MONTAGU, working in Belgium (HERRERA-ESTRELLA et al., 1983), had produced tobacco plants that were resistant to kanamycin and to methotrexate, a drug used to treat cancer and rheumatoid arthritis (SCHELL et al., 1983). R. FRALEY, S. ROGERS, and R. HORSCH at Monsanto had produced *Petunia* plants that were resistant to kanamycin (FRALEY et al., 1983). The Wisconsin group, headed by J. KEMP and T. HALL, had inserted a bean gene into a sunflower plant (MURAI and KEMP, 1982).

The major technical advance was the demonstration that transgenic plants could be regenerated from leaf discs following cocultivation with *Agrobacterium.* Subsequently transgenic plants have been produced from many families using this approach or modifications of it. Many other techniques have been tested for direct gene transfer in protoplasts. Electroporation was one of them (i.e., the application of high-voltage electric pulses to cells to induce transient membrane pores, allowing entry of macromolecules including DNA). For the first time, FROMM, TAYLOR, and WALBOT (1985) were successful in maize, rapeseed, soybean, rice, tobacco, and so on. The development of *Agrobacterium* as a vehicle for routine genetic transformation of plants was limited because of host-range limitations and difficulties with regeneration from protoplasts. POWELL-ABEL et al. (1986) developed the first transgenic plants using cDNA of coat protein gene of tobacco mosaic virus (TMV). In the same year, CROSSWAY et al. published the transformation of tobacco protoplasts by direct DNA microinjection, and the first field trials of transgenic plants were initiated.

In this context, bombardment of intact plant cells and tissues with high-velocity, DNA-coated microprojectiles, was considered crude but effective (KLEIN et al., 1987). Already in 1966, plant virologists MACKENZIE et al. had employed this technique to wound plant cells and facilitate entry of viral particles or nucleic acids. Today, about eight methods of gene transfer are applied, more or less efficiently, depending on the recipient crop and the target genes available:

1. By means of *Agrobacterium* and cocultivation
2. By means of viruses
3. By means of a particle gun
4. Via microinjection
5. By means of electroporation
6. By utilization of micro LASER beams
7. Via chimeric DNA/RNA plasty
8. With the help of liposomes

As targets of DNA introduction intact tissue, cell cultures, protoplasts, pollen cells, ovules, or chloroplasts are utilized. Input traits, such as resistance to herbicides, insects, viruses, fungi, bacteria, nematodes, or abiotic factors (see Chapter 4.2.2.1) and output traits, such as production of antibiotics, sera, immune bodies, metabolites, polymers, etc. (see Chapter 4.2.2.2), are differentiated.

The early transgenic plants were laboratory specimens, but subsequent research has developed transgenic plants with commercially useful traits such as resistance to herbicides, insects, and viruses. With the advancement in DNA technology and understanding of the structure and function of genes and their products, more and more transgenics are being developed in various research programs all over the world.

The ability to transfer new, alien, or artificial genes into old crop varieties has led to imaginative flights of fancy: a new range of disease- and stress-resistant plants, nitrogen fixation of nonlegumes, and amino-acid-balanced plant proteins. The same techniques can also be applied to improve modern varieties by old cultivars. For example, an ancestral emmer wheat allele encoding an *N*-acetylcysteine (NAC) transcription factor that accelerates senescence and increases nutrient remobilization from leaves to developing grains was isolated by positional cloning and transferred to advanced wheat varieties. The resulting wheat showed increased grain protein, zinc, and iron contents, which enhances the nutritional value as well as improves human nutrition and health (UAUY et al., 2006).

The concept of improving agriculture in the traditional sense by recombinant DNA technology became a reality with two dramatic inventions. One was that soybeans could be induced to be resistant to the nonselective, environmentally benign herbicide glyphosate (Roundup). The other was that the gene *Bt*, from the bacteria *Bacillus thuringensis,* could be transferred to the crop plant. *Bacillus thuringiensis* was used as an insecticide in the 1950s. The first gene encoding the *Bt* toxin was cloned in 1981. The regulation of *Bt* gene was known in 1986 and inserted into maize in 1990. *Bt* hybrids were first sold in 1997.

The creation of "Roundup-Ready" soybeans was to have an extremely rapid rate of adoption, unsurpassed in agriculture. *Bt* cotton was also rapidly adopted, but *Bt* maize was not widely adopted because the cost–benefit ratio was not as high as the maize rootworm incidence varied with location. By 1999, herbicide-resistant soybean accounted for 57 percent of the crop area, *Bt* cotton 55 percent, and *Bt* maize 22 percent in the United States.

Many traits controlled by transgenes are gain-of-function traits. This type of trait provides the added value expected from improved cultivars. Strong promoters controlling constitutive expression, such as cauliflower mosaic virus 35S promoter, currently drive these transgenes.

It should be noted a little critically that this process of alien gene transfer needed more than forty-five years. Compared with conventional breeding, it took much more time and money than any other new variety of maize. The story, beyond this point, is that agriculturally useful genes are not in surplus and their expression within foreign genomes is widely unresolved, although several positive and promising results were demonstrated.

The production of transgenic plants has to be seen in connection with traditional plant breeding, where humans, since prehistoric times, have selectively bred particular wild plants with good characteristics. Qualities, such as strength, yield, resistance against noxious organisms, and the ability to withstand wind and weather were improved by crossing the best individuals with each other. It still takes ten to fifteen years to develop a new variety using traditional breeding methods. Gene transfer techniques can reduce this time by a half and make it possible to selectively transfer genes so that it is possible to know exactly which characteristics have been introduced. It also gives the potential to introduce genes from nonrelated species or artificial genes.

Over the years, the boon of biotechnology has been confined to few developed countries, and developing countries are yet to harvest the benefits of modern technology. Australia was the first country in the world to have engineered and released a microorganism for biological control of crown gall in plants. Australia's first field trial of transgenic plants took place in 1991, involving transgenic potato varieties resistant to leafroll virus, developed by the Commonwealth Scientific and Industrial Research Organization (CSIRO, Australia), Division of Plant Industry. By 1994 the first genetically engineered crop plant was approved for commercial marketing—the "Flavr-Savr" tomato—designed to slow fruit ripening and increase shelf life (LEVETIN and McMAHON, 1996). During 1988-1991, several laboratories in Japan, using electroporation or *Ti* or *Ri*, reported successful production of transgenics in *Oryza saliva, Citrus sinensis, Cucumis melo, Lactuca saliva, Solanum tuberosum, Nicotiana tabacum, Morus alba, Actinidia chinensis, Atropa belledona, Brassica oleracea, Lycopersicon esculentum, Vigna angularis,* and *Vitis vinifera* (NAKAJIMA, 1991). For effective gene expression, promoter regions/genes have been identified for tissue-, age-, and pathogen-specific expression of the genes under transfer. Stable and useful transgenics for protein quality and for viral resistance have been produced in rice, potato, tomato, melon, and tobacco as well.

During the nine-year period 1996 to 2005, according to International Service for the Acquisition of Agri-Biotech Applications (ISAAA), global area of transgenic crops increased more than fortyfold, from 1.7 million ha in 1996 to 67.7 million ha in 2003. More than one quarter of the global transgenic crop area of 52.6 million ha in 2001 was grown in developing countries, where growth continued to be strong (Table 4.2).

In 2004, five principal countries grew 99 percent of the global transgenic crop area: the United States grew 47.6 million ha, Argentina 16.2 million ha, Canada 5.4 million ha, Brazil 5.0 million ha, and China 3.7 million ha. China had the highest year-on-year growth with a tripling of its *Bt* cotton area in 2004. The other eight countries that grew GM crops in 2005 were Mexico, Bulgaria, Uruguay, Romania, Spain, Indonesia, Kenya, and Germany. Indonesia reported commercializing a transgenic crop. Globally, the principal GM crops were soybean occupying 48.4 million ha in 2004 (60 percent of global area), followed by maize at 19.3 million ha (24 percent), cotton at 9.0 million ha (11 percent), and rapeseed at 4.3 million ha (5 percent). During the nine-year period 1996 to 2005, herbicide tolerance consistently was the dominant trait, with insect resistance being second. There is a cautious optimism that global area, and the number of farmers planting GM crops will continue to grow in 2005.

According to industry experts, while the United States will continue to account for the largest seed sales, China, India, and Brazil will present fast growth opportunities. AVENTIS CROPSCIENCE,

TABLE 4.2. World acreage of transgenic crop plants and countries involved.

Year	Area (million ha)	Countries
1996	1.7	6
1997	12.7	7
1998	29.4	8
1999	39.8	12
2000	44.2	13
2001	52.6	13
2002	58.7	16
2003	67.7	18
2004	81.0	19
2005	93.2	21

Source: Adapted from "Global Status of Commercialized. Transgenic Crops" ISAAA, 2005.

CARGILL, DELTA, KWS, PINE LAND, DUPONT, LIMAGRAIN MONSANTO, CIBA-GEIGY, BAYER, UNILEVER, ZENECA, and SYNGENTA are the leading global players in the GM crops segment. They cover a wide range of products, which include maize, soybean, cotton, tomato, potato alfalfa, petunia, rape/mustard, rice, wheat, mustard, beet, barley, gram, cabbage, papaya, and tobacco.

4.2.2.1. Future Transgenic Crop Plants and Genetic Engineering

In order to effect improvement in any plant's productivity or performance, the breeder must be fully aware of the attributes and breeding history of the plant in question. In addition to more conventional concerns, such as disease resistance and day-length control of flowering, some destabilized environmental factors must now be given more attention. The predominant genotypes for each crop may in the future need to be reassessed and modified in light of their responses to increasing mean global atmospheric CO_2 concentration, climatic changes, or increasing penetration of damaging UV light and deficient stratospheric ozone. Within the next ten or twenty years, a few research areas may become very important for crop improvement:

- *Increased yield and increased reliability of performance:* The increase of leaf photosynthetic rates seems to be a straightforward way. Conventional breeding already demonstrated genotypes with superior photosynthetic rates in maize, wheat, and soybean. One current target for molecular modification of photosynthesis is to introduce the precursor pathway for organic acid fixation of CO_2 (C4 pathway) into C3 species. Transgenic lines of rice have already been developed that have higher levels of phosphoenolpyruvate carboxylase activity, but their leaf photosynthetic rates are still comparable to nontransformed lines.
- *Changes in plant architecture modifying balanced proportions of tuber, seed, leaves, or internal characters.*
- *Changes in pest and disease resistance:* Although crop plants belong to certain plant families, most of their diseases are specific to a particular species. Therefore, resistance genes are often not usable across the genus barrier despite the crossability and in-

compatibility problems. However, recent transgenic methods demonstrate the feasibility of transferring nonhost resistance genes between distantly related grasses to control specific diseases. For example, a maize resistance gene, *Rxo1*, that recognizes a rice pathogen, *Xanthomonas oryzae* pv. *oryzicola,* which causes bacterial streak disease, conditioned a resistance reaction to a diverse collection of pathogen strains of rice (ZHAO et al., 2005).

- *Improved tolerance to abiotic stress:* The increase of growth rate of individual seeds seems to be one approach. Transformation of ADP-glucose pyrophosphorylase in seed results in less sensitivity to phosphorus inhibition of wheat and rice. Inadequate water availability is another crucial limitation to crop yield in most environments and has been the focus for genetic improvement of crops for many years. The ratio between plant mass accumulation and transpiration, i.e., water-use efficiency, has limited flexibility owing to the physics and physiology of leaf gas exchange. One molecular approach that has received considerable attention is the possibility that solute accumulation in plants, or osmoprotection, might confer drought tolerance (SINCLAIR, PURCELL, and SNELLER, 2004). Transformations are being attempted for osmolyte accumulation and increased production of compatible solutes under water-deficit conditions.
- *Apomixis to fix hybrid vigor.*
- *Male sterility systems with transgenics for hybrid seed in self-pollinating crops.*
- *Parthenocarpy for seedless vegetables and fruit trees:* Plants capable of forming fruits without fertilization can increase fruit acceptance by consumers. To achieve parthenocarpic development, it is common practice to treat flower buds with synthetic auxins. To mimic the hormonal effects by genetic engineering, the expression of a gene able to increase auxin content and activity should be induced in the ovule. In some experiments, the gene *iaaM* from *Pseudomonas syringae* under the control of an ovule-specific promoter from *Anthirrhinum majus* was utilized to induce parthenocarpic development in transgenic tobacco and eggplants. This approach could also be valuable in other horticultural crops, such as pepper, tomatoes, melons, etc.

- *Short-cycling for rapid improvement of forest and fruit trees as well as tuber crops:* The ability to control sprouting time in potato tubers is of considerable economic importance. In order to prevent chemical treatment for stimulating tuber control, the transformation of the pyrophosphatase gene from *Escherichia coli* under the control of the tuber-specific patatin promoter was realized. Transformed potatoes displayed a significantly accelerated sprouting by six to seven weeks. In citrus trees, the generation time was reduced. Citrus trees have a long juvenile phase that delays their reproductive development by between six and twenty years. Therefore, juvenile citrus seedlings were transformed with *A. thaliana Lfy* and *Ap1* genes, which promote flower initiation. Both types of transformed seedlings produced fertile flowers as early as the first year. These traits are submitted to the offspring as dominant alleles, generating trees with a generation time of one year from seed to seed. Constitutive expression of the *Lfy* gene also promoted flower initiation in transgenic rice.

- *Nutritional and micronutritional efficiency of cereal and tuber crops:* Increased nitrogen accumulation by crops has been a crucial feature of past yield increases. Increased nitrogen accumulation has usually resulted from applications of nitrogen to the soil and genetic improvement of plants to accumulate and store greater quantities. The goal of molecular changes is to increase nitrogen-use efficiency. The activities of specific enzymes involved in nitrogen metabolism have been targeted for transformation, e.g., the overexpression of a glutamine synthetase gene. The detrimental effects of salt on plants are a consequence of both a water deficit and the effects of excess sodium ions on key biochemical processes. To tolerate high levels of salts, plants should be able to use ions for osmotic adjustment and internally to distribute these ions to keep sodium away from the cytosol. The first transgenic approach introduced genes that modulated cation transport systems. Hence, transgenic tomato plants overexpressing a vacuolar Na^+/H^+ antiport were able to grow, flower, and produce fruits in the presence of 200 mM sodium chloride. Another overexpression of the *Hal1* gene from yeast in transgenic tomato plants had also a positive effect on salt tolerance by reducing K^+ loss and decreasing intracellular Na^+ from the cells under salt stress (HELLER, 2003). In arid and semiarid regions of the world, the soils are alkaline in nature, and therefore crop yields are limited by the lack of available iron (cf. Chapter 3.7.4). Under iron stress, some plants release specific Fe(III)-binding compounds, known as siderophores, which bind the otherwise insoluble Fe(III) and transport it to the root surface. To increase the quantity of siderophores released under conditions of low iron availability, two barley genes coding for the enzyme nicotianamine aminotransferase together with the endogenous promoters were introduced into rice plants. Transformed rice plants withstand iron deprivation remarkably well, resulting in a fourfold increase in grain yield as compared with the control plants. Alternately, increasing the rate-limiting step of Fe(III) chelate reduction, which reduces iron to the more soluble Fe(II) form, might enhance iron uptake in alkaline soils. In 2004, "HarvestPlus" was implemented by the Consultative Group on International Agricultural Research (CGIAR) Global Challenge Programs for breeding crops with better nutritional value. Initially, six crops and three nutrients were targeted (beans, cassava, maize, wheat, rice, and sweet potato).

- *Converting annual into perennial crops for sustainable agricultural systems.*
- *DNA repair in plants:* The expected increase in UV-B radiation at the earth's surface by the early 21st century is of considerable concern because UV causes damage to genetic material (DNA). In the case of crop plants, unrepaired DNA damage may have penalty in yield losses. From *Arabidopsis* a gene coding for nucleotide repair endonuclease, ERCC1, was identified and isolated; its protein plays a pivotal role in the DNA nucleotide excision repair process. It is the first plant gene involved in DNA excision repair so far reported (XU et al., 1998, 1999).

Banning transgenic crops in the farming system will be ineffective because the potential benefits are so great. Whatever scientists do to develop crops that eliminate or reduce the utilization of polluting agro-chemicals in the farming systems must be welcomed by farmers and consumers. For example, one interesting approach for developing resistant transgenic crops may be through the improvement of the plant's own defense system. Inducible and tissue-specific promoters could assist in this endeavor.

Collective approval may lead to new partnerships, cooperation, or joint ventures in research and development between scientists in the public and private sectors, which will benefit farmers and consumers with profits and high-quality products, respectively. Any potential risk in human development associated with biotechnology applications in agriculture will be easily resolved in a democratic society. The public needs to choose between being safely self-regulated or to follow safety regulations as agreed by lawmakers, after listening to the views of scientists, producers, and consumers.

In 2005, a study showed no maize, soybeans, or rapeseed crop in the United States that did not have traces of genetically engineered counterparts. An environmental advocacy group tested nontransgenic seed from major seed suppliers and found low but detectable levels of DNA from two transgenic varieties in 50 and 83 percent of maize and soybean respectively and between 83 and 100 percent of rapeseed. Eight percent of U.S. soybeans and 38 percent of the maize crop are genetically engineered to be herbicide-resistant, to manufacture their own pesticides, or both. However, such mixing is an economic threat to organic growers, who risk losing customers if their products are not 100 percent nontransgenic. Europe and Japan have long refused to purchase transgenic products.

The risk is manifold: Extensive use of the same herbicide for a long time in many different crops may hasten the appearance of tolerant weeds, and the herbicide-tolerant gene may introgress into a wild or weed wild relative of the transgenic crop. MIKKELSON, ANDERSON, and JØRGENSEN (1996) have demonstrated the introgression of the BASTA tolerant gene from transgenic rapeseed *(Brassica napus)* to a weed *(B. campestris)*. Transgenic crops could decrease the biodiversity. Concerning the latter, some contrary examples are already in the fields:

1. Herbicide-resistant transgenic crops have made new crop rotations possible, owing to more flexible weed control and less persistent herbicides (e.g., grain legumes in Canadian rapeseed rotations).

2. More birds and beneficial insects are seen in GM cotton fields.
3. Still in the research phase is the use of transgenic fungi to slow down the devastating disease that has almost eliminated the native chestnut tree from North American forests.
4. A long-run global impact of biotechnology may be increased space for wilderness, since the greatest threat to the earth's biodiversity is habitat loss through conversion of natural ecosystems to agriculture. Conservation of biodiversity in situ will continue to be possible if high-yield practices allow the same amount of land to support a higher standard of living, more people, or both. The deployment of transgenic crops could dramatically reduce use of fungicides and pesticides that are still more heavily used in Europe than in North America.

The public should see biotechnology as a safe tool for scientific crop improvement, because it helps in the fight against hunger and poverty. Therefore, research funding should be allocated accordingly to long-term plant breeding programs, which include biotechnology as one of its tools. In this way, we may effectively face the serious challenge of feeding the rapidly growing world population in the next millennium. However, alternative interests should not be ignored. The European Consortium for Organic Plant Breeding (ECO-PB), founded in 2001 at Driebergen (the Netherlands), is one of them. It provides a platform for discussion and exchange of knowledge and experiences, international support of organic plant breeding programs, and the development of scientific concepts of an organic plant breeding.

4.2.2.2. "Farmerceuticals" and Other Exotic Characters of Modern Crop Plants

Modern crops are increasingly used for various industrial applications. They contain specific raw materials like fibers, oils, and starches for industrial products. Tailor-made plant compounds for specific industrial processes are a huge challenge and may contribute to a more sustainable production. In the near future, plants might be used as factories for vaccines and pharmaceuticals (see Table 4.3). The focus is on diseases that are more or less eradicated in industrialized nations but are hard to combat in developing countries because of the lack of money and infrastructure. Field experiments have already been car-

TABLE 4.3. Renewable crops for better nutritional value and future production of pharmaceuticals or other drugs.

Crop	Trait	Expected benefit
Soybean, rapeseed	Improved fatty acid; amino acid composition; reduced phytate content	Heart disease prevention; increased nutritional value for animals
Tomato	Delayed ripening	Increased shelf life
Strawberry	Freeze-thaw tolerance	Better taste
Cassava	Improved protein content, elimination of cyanide	Increased nutritional value, reduced toxicity
Potato	Accelerated starch synthesis; multi-component vaccine; spider silk production	Better digestibility; human immunization; artificial fibers with novel characteristics
Soybean	Increased sweetness	Sweeter taste
Rice	Provitamin A and iron content; elimination of allergens	Increased nutritional value; allergy prevention
Cereals (wheat, rye, and oat)	Gluten removal	Allergy prevention
Maize	Avidin, bovine trypsin	
Banana	Cholera vaccine	Drug production

Source: Economic Research Service, USDA, and International Food Information Council.

ried out on potatoes, bananas, and pineapples containing cholera, *E. coli,* and hepatitis B vaccines.

Researchers have just succeeded in immunizing mice with transgenic potatoes. As cereal grains are deficient in certain essential micronutrients, including iron or zinc, several approaches have been used to increase accumulation (e.g., alter iron metabolism). Since ferritin is a general iron storage protein in all living organisms, ferritin genes have been introduced into rice and wheat plants. Transgenic rice plants can express soybean and common bean ferritin under the control of the seed-specific rice Glu-B1 promoter. They accumulate up to three times more iron as compared with wild-type seeds.

In order to increase iron accumulation and improve its absorption in the human intestine, two approaches have been adopted. Either the level of the main inhibitor of iron absorption, phytic acid, is decreased by introduction of a heat-tolerant phytase from *Aspergillus fumigatus,* or a cystein-rich metallothionein-like protein gene (*rgMt*) is overexpressed in the recipient plant. It can be expected that high-phytase rice, with increased iron content and rich in cysteine-peptide, greatly improves iron supply in rice-consuming populations.

In addition, the amino acid composition of crop plants can be modified. For example, potatoes contain limited amounts of the essential amino acids lysine, tryptophan, methionine, and cysteine. In order to improve the nutritional value of potatoes, a nonallergenic seed albumin gene *AmA1* from *Amaranthus hypochondriacus* was transferred to potato plants. The seed-specific albumin was under the control of a tuber-specific and a constitutive gene promoter, respectively. In transgenic lines, a 35 to 45 percent increase of total protein content was possible, which corresponds to an increase in most essential amino acids.

The reduction of antinutritive factors is as feasible. For instance, cassava is one of the few plants that contains toxic cyanogenic glycosides in the leaves and shoots. Therefore, the genes that are responsible for the production of the glycosides were isolated. Subsequently, cassava plants were transformed with antisense constructs of the respective genes *Cyp79D1* and *Cyp79D2*. Another strategy for reducing the cyanide toxicity is to introduce a gene that codes for the enzyme hydroxynitrile lyase. This enzyme breaks down the major cyanogen acetone cyanohydrin.

The development started in the 1980s. In 1983, the first industrial production of secondary metabolites by suspension cultures of *Lithospermum* spp. was carried out by MITSUI PETROCHEMICALS (Japan). The beneficial use of elicitors in cell suspension cultures was early recognized by B. WOLTERS and U. EILERT. The hairy root production of metabolites in *Hyoscyamus muticus* was demonstrated by H. E. FLORES and P. FILNER (1985). These roots produced more hyoscyamine than whole plants. The hairy root method provides several new approaches for breeding and biotechnology, such as better understanding and manipulating of root-specific metabolism, coculture with vesicular-arbuscular mycorrhizae (VAM) fungi to increase secondary metabolite production, coculture with insects to study pathogenesis, commercial exploitation of bioactive root exudates, or coculture with shooty teratomas to exploit both root- and shoot-based metabolism for biotransformation.

In 1986, it was shown by BARTA et al. that tobacco plants and sunflower calluses could express recombinant human growth hormone as a fusion protein. In 1989, the first plant-derived recombinant antibody, the full-size IgG, was produced in tobacco (HIATT, CAFFERKEY, and BOWDISH, 1989) and, in 1990, the first native human protein (serum albumin) in tobacco and potato (SIJMONS et al., 1990). The first plant-derived protein polymer (artificial elastin) could be harvested from recombinant tobacco (ZHANG, URRY, and DANIELL, 1996). Even more exciting is the production of spider silk proteins in transgenic tobacco and potato (SCHELLER et al., 2001). The expression level of best plants reached up to 4 percent of the total soluble protein. It was stable in tobacco leaves and in potato leaves and tubers.

"Green biotechnology" has been used to engineer plants that contain a gene derived from a human pathogen (TACKER et al., 1998). An antigenic protein encoded by this foreign DNA can accumulate in the resultant plant tissues. Results from preclinical trials showed that antigenic proteins harvested from transgenic plants were able to keep the immunogenic properties if purified. ARAKAWA et al. (1998) demonstrated the ability of transgenic food crops to induce protective immunity in mice against a bacterial enterotoxin, such as cholera toxin B subunit pentamer with affinity for ganglioside. In addition, potato tubers have been used successfully as a biofactory for high-level output of a recombinant single chain antibody (ARTSAENKO et al., 1998).

The other example of high provitamin A "golden rice" came to minds of YE et al. (2000). Deficiencies in dietary vitamin A are one of the leading causes of child blindness in the developing world. "Golden rice" was engineered to contain provitamin A based in the introduction of two genes from daffodil and one from the bacterium *Erwinia uredovora*. Common rice contains no provitamin A. Thus, the ability to introduce these genes to create a new biochemical pathway in rice is an exciting development. However, five years after initiating the project, the international organization Greenpeace claims that this project is a technical failure, not suited to overcome malnutrition and worse, and it is drawing funding and attention away from the real solutions to combat vitamin A deficiency.

Another example is transgenic tomato, which received the bacterial carotenoid gene *crt1* encoding for the phytoene desaturase and showed an increase of β-carotene content up to 45 percent of the total carotenoid content. In 2004, a production system for recombinant protein C5-1, a diagnostic antihuman IgC for phenotyping and cross-matching red blood cells from donors and recipients in blood banks, was developed based on the leaves of alfalfa *(Medicago sativa)*. The perennial alfalfa can be easily propagated by stem-cutting to create large populations. In greenhouses, these populations can be harvested ten times a year, and the recombinant plants can be maintained for more than five years for protein extraction.

KOIVU (2004) proposed a novel sprouting technology for recombinant protein production. Sprouted seeds or sprouts from oilseed rape *(Brassica napus)* are renowned for their excellent nutritional properties. They are often utilized as a protein-rich food and some beneficial phytochemicals, such as vitamins A, E, and C, which are antioxidants. In a contained system, transgenic seeds carrying the gene encoding for a specific protein of interest are first produced and harvested in a greenhouse and then sprouted in an airlift tank. The recombinant protein is produced during sprouting and extracted from the sprouts. Alternatively, the protein can be removed directly from the growth medium.

For industrial countries, a low-calorie sugar could be of interest. New sugar beet plants were engineered that produce fructan, a low-calorie sweetener, by inserting a single gene from Jerusalem artichoke *(Helianthus tuberosus)*. The gene encodes an enzyme for converting sucrose to fructan. Short-chain fructans have the same sweetness as sucrose but provide no calories, as humans lack the fructan-degrading enzymes necessary to digest them.

Vaccine production in plants is attempted for antiallergic medicals. Current allergy immunotherapy involving injections of a crude pollen (allergen) extract has the disadvantage that it can lead to severe and life-threatening anaphylactic side effects. Recently, nonallergenicorms of grass pollen allergens were produced via site-directed mutagenesis (see earlier). Low toxicity, solubility, and production of high amounts of correctly folded recombinant proteins at low cost are prerequisites for developing new vaccination strategies. Therefore, plant-based expression systems developed for modified (engineered) allergens are potentially suitable for immunotherapy (BHALLA and SINGH, 2004).

In 2005, an American research group was even able to produce a vaccine in transformed tobacco and tomato plants against the SARS virus, which, in 2002, was identified in China for the first time. Another American group was able to transfer a microbial gene, the superoxide reductase, to plants for future (Mars) space missions. W. BOSS and A. GRUNDEN from North Carolina State University, Raleigh, have combined beneficial characteristics from an extremophile sea-dwelling, single-celled organism, *Pyrococcus furiosus,* into model plants like tobacco and *Arabidopsis thaliana.* They hope to improve extreme-temperature protection of plants.

4.2.2.3. Prospects

In developing countries, peri-urban agriculture and home gardening are becoming more important for national food security because of rapid urban expansion. Hence, new cultivars will be needed to fit into intensive production systems, which may provide the food required to satisfy urban world demands of this century. Specific plant architecture, tolerance to urban pollution, efficient nutrient uptake, and crop acclimatization to new substrates for growing are, among others, the plant characteristics required for this kind of agriculture. Actually, the first step toward this new type of agronomy has already been taken. The production of rapeseed for biodiesel reached such a high level in several European countries that it became a particular branch of agriculture. Alcohol production from sugarcane in Brazil is another indication for structural changes in agriculture.

Food crops with low fats and high in specific amino acids may be another need to satisfy people who wish to change their eating habits. If genes controlling these characteristics do not exist in a specific crop pool or gene banks, they may be incorporated through genetic engineering.

Some publications anticipated that in the 21st century food would not need to be harvested from farmer's fields (ANDERSON, 1996). Tissue culture of certain parts of the plant may provide a means to achieve success in this endeavor. For example, edible portions of fruit crops could be grown in vitro.

A steady and cheap supply of these edible plant parts will be required in this new agri-business. It will take some time before such a process can be scaled up for commercial output. Nonetheless, a biotech company for producing a vanilla extract through cell culture got it patented (RAO and RAVISHANKAR, 2000). Of course, this technique will not replace farming. This biotechnique, as well as other new farming methods, offers a means for new ways of producing food, feed, or other products, such as medicines, solvents, dyes, and noncooking oils for many years. Hence, it would not be surprising to see, a few years from now, entire farms without food crops but growing transgenic plants to produce new products, e.g., edible plastic from peas or plant oils to manufacture hydraulic fluids and nylon (GRACE, 1997). This new rural activity may result in important changes in the national economic sector.

"Pharming" has been added to the dictionary of agricultural biotechnology, indicating a new kind of production system.

Chapter 5

Intellectual Property Rights, Plant Variety Protection, and Patenting

Intellectual property rights are legal and institutional devices to protect inventions, patents, plant breeders' rights, etc. They have never been more economically and politically important than they are in today's industrial societies, nor have they ever been so controversial. Patents provide inventors with legal rights to prevent others from using, selling, or importing their inventions for a fixed period, usually twenty years. In a modern sense, patenting of plants, seeds, and plant parts does not appear to go much further back than the 1930s.

Nevertheless, the first plant patent seems to have been granted to red currant *(Ribes rubrum)* in 1621, when in England Sir Francis BACON (scientist, philosopher, and Lord Chancellor of Her Majesty) confirmed a patent privilege to the English Crown to trade with currant, salt, starch, herrings, vinegar, etc. In 1716, a further patent was granted in England for squeezing oil from sunflower seed, and in 1724, T. GREENING was granted a patent for "grafting or budding the English elm upon the stock of the Dutch elm." In 1785, P. le BROCQ from Jersey (Channel Islands) acquired a patent for "rearing, cultivating, training, and bringing to perfection, all kinds of fruit trees, shrubs, and plants; protecting their leaves, blossoms, flowers and fruits."

However, the history of intellectual property rights for plants is mainly connected with the development of breeding institutions during the past two centuries (see Chapter 3.2). In 1895, the first "law over the forgery of agricultural products" (Seed law) became effec-

Concise Encyclopedia of Crop Improvement
© 2007 by The Haworth Press, Inc. All rights reserved.
doi:10.1300/5891_05

tive in the Austrian-Hungarian monarchy. In Germany, registered breeding stations and elite breeds were kept beginning in 1888. In 1896, the German Agricultural Society (DLG, or Deutsche Landwirtschaftsgesellschaft) established the first "Rules of Seed Registering." It led to the elite register of 1905, which also contained regulations for variety testing and purity.

Before, in 1869, F. NOBBE, Head of the Plant Physiology Experiment Tharandt (Saxonia, Germany), established the first official and independent seed control system worldwide. Seed morphology, purity, and germability were investigated and certified. Regulations were described in the Statut Betreffend die CONTROLE Landwirtschaftlicher SAATWAAREN (Statute Concerning Control of Agricultural Seedware). Soon after, similar testing institutes followed in Denmark, Austria, Hungary, Belgium, Italy, and the United States. NOBBE's *Handbuch der Samenkunde* (Handbook of Seed Science; 1876) was a standard textbook at the time.

In the United States, land-grant institutions were established in 1860s. The United States Department of Agriculture was founded in 1862, the same year as the passage of the Morrill Land-Grant Act, which established most of the country's agricultural colleges. In the 1870s to 1880s, the federal government began to markedly increase its commitments to agricultural research. However, this situation discouraged private investments in plant breeding because it was difficult to maintain control over sales and markets and recoup investments. Particularly, with the introduction of hybrid maize in the United States during the 1930s, saved seed was no longer an option. The hybrid varieties could be protected through trade secrecy. Proprietary maize hybrids were initially based on public inbreds. The pressures from investors to develop and support commercial plant breeding increased. However, all released varieties and hybrids could be used as breeding materials. During this period, trade secrets and contracts were often used as low-cost alternatives to more formal means of protection. Prior to 1930, farmers still had direct access to seed and germplasm since most crops were "true breeding" and seed was easily saved. Moreover, most breeding was publicly-financed. However, the use of and access to breeding strains, germplasm, and varieties was limited or restricted, even during cooperative testing. In 1930, the Plant Patent Act was the first legislation in the United States to protect horticultural plants and nursery stocks of asexually propagated crops. Potato was still excluded. As a critierion for differentiation, the variety must be "distinct" and "new." The control was administered through the U.S. Patent Office.

In Europe, patents were occasionally granted around this time. In Germany, W. LAUBE (1892-1963), the breeder of the first tetraploid rye variety "Petkuser Tetraroggen" (1951), initiated a similar law ("Sortenschutzgesetz") protecting breeders from plagiarism. It was passed in 1932. Later, the German Appeal Board accepted an application for a patent that claimed seed material of certain varieties of lupin (see Chapter 3.5). Nevertheless, the U.S. Plant Patent Act stimulated one of the European initiatives leading, on November 17, 1938, to the foundation of ASSINSEL (Association Internationale des Sélectionneurs pour la Protection des Obtentions Végétales) at Amsterdam (the Netherlands) with Denmark, France, Netherlands, and Germany as the first member countries. During the 1940s and early 1950s, countries adopted a range of approaches, from denying all intellectual property protection (e.g., United Kingdom and Denmark) to allowing patents (Italy from 1948, France from 1949, Belgium from 1950) and creating specific intellectual property systems for plant varieties (e.g., the Netherlands in 1941, Austria in 1946, and Germany in 1953). In 1952, South Africa introduced a similar but modified patent system for plants as in the United States.

In 1970, the American Plant Variety Protection Act (PVPA) was passed. Sexually propagated varieties could be protected. The primary goal was to promote commercial investments in plant breeding and to provide a "patent-like" protection for plants reproduced by seed. The right of owners was still limited. The protection included a period of twenty years of exclusive rights for the breeder considering entire plants and harvested material. The right was issued by the USDA. Seed sale was only allowed through authorized dealers.

In 1961, in Europe a similar system was put into law by "Union pour la Protection des Obtentions Vegetales (UPOV)." Member countries were Belgium, Denmark, Germany, France, Great Britain, Italy, the Netherlands, and Switzerland. After this, breeders have had the right to exploit products of their profession. It included authorization for plant production, sale, and marketing. The variety proposed must be "distinct, uniform, stable," and novel in at least one trait. However, problems remained, such as widespread "brown-bagging" (i.e., illegal sales and use), erratic and inadequate enforcement, enforcement responsibility of patent holder, only minor penalties for violation or infringement, and concern over impact on restriction of germplasm exchange and crop diversity.

This law led to an increase in private breeding, but only for a few crops. Market size and profit margins were primary determinants of commercial success. Crops, such as autogamous wheat and barley, faced with low profit margins for seed and extensive pirating, received only limited private investments. By this time, many government-owned plant-breeding institutes had been privatized, and usually large chemical corporations had bought them. Many of these corporations have also been buying seed production and distribution organizations. This process of reorganization, concentration, and monopolization of breeding is still ongoing.

In 1980, the BAYH-DOLE Act and, in 1986, the Technology Transfer Act had an additional impact on public plant breeding and research in the United States. It established that universities had the right to obtain patents and commercialize inventions created under government grants, that licenses and royalties could be sought by universities as a means to generate revenue as government support declined, and that, as an inventor, a university employee has the right to receive a portion of the royalties on the invention.

The Canadian Plant Breeders' Rights Act (PBR Act) came into force on August 1, 1990. This PBR Act is administered by the Plant Breeders' Rights Office, which is part of the Canadian Food Inspection Agency. The PBR Act allows the developers of new varieties to recover their investment in research and development by giving them control over the multiplication and sale of the reproductive material of a new variety. The rights also include the ability to charge a royalty. In order to receive a grant of rights, varieties must be new, distinct, uniform, and stable. Two notable exceptions to a holder's rights are that protected varieties may be used for breeding and developing new plant varieties and that farmers may save and use their own seed of protected varieties without infringing on the holder's rights. This second exception is referred to as Farmers' Privilege.

In 1994, an amendment of American PVPA eliminated the "saved-seed" provision; i.e., the farmer can save seed only for on-farm replanting. This amendment brought the PVPA into accordance with UPOV in Europe. In the same year, the European Community (EC) in Brussels (Belgium) passed a similar law (2100/94) that unified the variety protection Europe-wide.

In 2001, plant patenting reaffirmed that utility patents, plant patents, and plant variety protection are different, but must be treated as complementary (BARTON and BERGER, 2001).

At present, in the United States, intellectual property protection for plants is provided through plant patents, plant variety protection, and utility patents. Plant patents provide protection for asexually reproduced varieties excluding tubers. Plant variety protection provides protection for sexually reproduced varieties including tubers, F1 hybrids, and essentially derived varieties. Utility patents currently offer protection for any plant type or plant parts. A plant variety can also receive double protection under a utility patent and plant variety protection.

On the other hand, European governments defied the EC Commission in 2005 by voting for the right to keep bans on patented, i.e., genetically modified crops and food. Five member states—Austria, Luxembourg, Germany, France, and Greece—were under pressure to give up their current bans because of a trade dispute in which the United States claimed the bans were illegal. The bans were imposed in the five countries between 1997 and 2000 for safety reasons.

In 1980, the first utility patent of "living organisms" was issued in the United States. It was a landmark decision of the U.S. Supreme Court in the case of *Diamond v. Chakrabarty* including the U.S. Patent and Trademark Office. It broadens the patent law to encompass living organisms and established that anything made by man is patentable, i.e., ownership of plant varieties, traits, parts, and processes. The claims can be broad-based, including entire species, plant parts, seeds, cell cultures, plant tissues, transformed cells, expressed proteins, threshold traits, and genes themselves. The standards for issuance of a utility patent are the novel character in relation to "prior art," the usefulness, and the obviously innovative step. Thus, a utility patent provides more intellectual property rights than a plant variety patent, but at a higher cost and standard for issuance. It allows prohibition of farm-saved seed and prohibition of use in breeding.

In this context the so-called "terminator technology" should be mentioned. Two methods are available: (1) "Traitor"—officially known as Trait-specific Genetic Use Restriction Technology (T-GURT)—incorporates a control mechanism that requires yearly applications of a proprietary chemical to activate desirable traits in the crop. The farmer can save and replant seeds but cannot gain the benefits of the controlled traits unless he pays for the activating chemical each year; (2) "Terminator"—officially named the Technology Protection System (TPS)—incorporates a trait that kills developing plant embryos, so seeds cannot be saved and replanted in subsequent years. Both methods avoid the difficulties associated with enforcing "no replanting" agreements and ensure the seed companies the investments in their new varieties.

The first Utility Patent for genetically engineered maize was granted in 1985, also in the United States. The chief interest in utility patents came from inventors in biotechnology products and processes. However, seed companies have looked toward utility patents for additional protection beyond that afforded by Plant Variety Protection. By 1988 over forty patents on crop plants had been issued. To date there are more than 2,000 U.S. patents with claims to plants, seeds, or plant parts.

Concerns remain about plant patents, despite restrictions placed on patented varieties for subsequent use in breeding, of potential negative impact on crop diversity, and of increased domination of breeding by larger companies. In the course of the last three generations, about 75 percent of crop and horticultural plants vanished from fields and gardens! One study estimates that 34,000 species of plants (~12.5 percent of the world's flora) are facing extinction. Another Food and Agricultural Organization (FAO) report (ANONYMOUS, 1998) stated, for instance, that in China of the almost 10,000 wheat varieties in use in 1949, only 1,000 remained in the 1970s; and in the United States 95 percent of the cabbage, 91 percent of the maize, 94 percent of the pea, and 81 percent of the tomato varieties cultivated in the 19th century have been lost.

Modern plant breeding, whether classical or through genetic engineering, comes with issues of concern, particularly with regard to food crops. The question of whether breeding can have a negative effect on nutritional value is a part of the declining genetic crop and genotype diversity. By some reports, the majority of the indigenous landraces that have developed since humans began to cultivate the environment have been lost because of disuse or neglect. During the past century, therefore, various national governments and international organizations developed a range of programs and institutes to identify and preserve plant genetic resources either in situ or ex situ. While these collections are now quite large, with more than 7 million discrete accessions globally (in ~1,300 genebanks of nearly eighty countries), it is unclear whether the agronomic vitality or traditional knowledge surrounding plant genetic resources are being preserved. Moreover, there is significant debate about who owns the rights to determine access and use of the materials and who should benefit from any discoveries or inventions using those plant materials, particularly the development of new biotechnological tools, which has lowered the costs and widened the possibilities of finding economically valuable traits in indigenous plants.

Although relatively little direct research in this area has been done, there are scientific indications that, by favoring certain aspects of a plant's development and marketing, other aspects may be retarded. Nutritional analysis of vegetables of the United States done in 1950 and in 1999 found substantial decreases in six of thirteen nutrients measured, including 6 percent of protein and 38 percent of riboflavin. Reductions in calcium, phosphorus, iron, and ascorbic acid were also found. It is explained by reduction and changes in cultivated varieties between 1950 and 1999.

Chapter 6

In the Service of CERES—A Gallery of Breeders, Geneticists, and Persons Associated with Crop Improvement and Plant Breeding

AARONSOHN, A. (1876-1919): well known as a botanical explorer and an agricultural expert on semiarid regions of the Mediterranean basin; he was born in Romania and moved to Haifa (Palestine) in 1882; he studied gardening and agriculture in France, and rediscovered wild emmer, the tetraploid ancestor of durum and bread wheats; he suggested that wild emmer wheat, as well as other wild forms of domesticated plants, may serve as "gene pools": he formulated a vision for improvement of agricultural varieties and practices, particularly aimed at dry-land farming.

ACHARD, F. C. (1753-1821): German scientist and chemist; the first person who selected beet plants successfully for sugar production; he is considered to be the founder of the sugar beet industry; by crossing the "Weisse Mangoldrübe" (Zuckerwurzel = sugar root) and the "Roter Mangold" (Rübenmangold or Futterrübe = fodder beet), he was able to select genotypes showing high sugar content, marking the birth of a new crop plant, the sugar beet; the first name for the sugar beet was "Schlesische Rübe"; first production of sugar beets started in Central Germany around the Magdeburg–Halberstadt–Kleinwanzleben region, where heavy, very fertile loess soil was available (it is still the center of sugar beet production of Germany);

Concise Encyclopedia of Crop Improvement
© 2007 by The Haworth Press, Inc. All rights reserved.
doi:10.1300/5891_06

later he founded the first sugar beet factory of the world at Cunnern (Silesian, Germany/ Poland).

ÅKERBERG, E. (1906-1991): honorary member of the Swedish Seed Association and director of the Association from 1956 to 1971; his name was synonymous with Svalöf and its plant breeding and research during this fifteen-year period; he grew up on an agricultural experimental station, Flahult, Småland (Sweden), where his father was the farm manager; he was trained as an agronomist at Alnarp and qualified in 1931; parallel with his studies at Alnarp, he was a student of Science at Lund University; he specialized in botany and genetics while earning his BSc degree in 1928; while continuing his postgraduate studies at Lund, he took up a position as plant breeder at the Weibullsholm plant-breeding station, Landskrona (Sweden), during 1932-1938; he bred many crops such as oats, peas and beans, clover and grasses, and potatoes; his PhD thesis work was devoted to an analysis of the grass species *Poa pratensis* and *Poa alpina* and crosses between them; in 1938, when he left Weibullsholm to become the head at one of the Svalöf branch stations in Norrland (Lännäs), he was able to present a new red clover variety, "Resistenta"; after six years of breeding work at Lännäs, he moved to Uppsala as head of the Swedish Seed Association's Ultuna branch station; in early 1956, ÅKERBERG moved to Svalöf as the new director of the Swedish Seed Association; he succeeded >>> A. ÅKERMAN; in the same year, 1956, he joined a group of leading European plant-breeding professors who started the European Plant Breeders Union (EUCARPIA); he was the president of EUCARPIA during 1965-1968; later, ÅKERBERG became vice-chairman of the Swedish State Agricultural Research Council; in finding new alternatives in crop production, he introduced rape as a new silage crop in northern Sweden and *Vicia faba* as another seed crop for the southern part of the country.

ÅKERMANN, A. (1887-1955): a Swedish botanist and breeder working at Svalöf; in 1904, under the guidance of Prof. H. NILSSON-EHLE, he started genetic studies on quantitative traits of wheat and oats; during the 1920s he carried out experiments on cold tolerance in wheats and other cereals; in the 1930s he began studies on the baking quality of wheat; in 1931, he established the first Swedish laboratory for exploring the baking and oil quality of wheat and/or rapeseed; his wheat varieties, derived from Swedish landraces and English square-head strains, contributed to a yield increase in Sweden of 50 percent during the first half of the 20th century.

ALLARD, R. W. (1919-2003): an American breeder born in the San Fernando Valley; by the age of ten, he had decided he wanted to be a plant breeder; he graduated from the College of Agriculture at University of California, Davis; he earned his doctorate from the University of Wisconsin at Madison; after 1945, he started his academic career at the University of California, where he served until his retirement; he strongly contributed to plant population genetics, predominantly self-fertilizing plants; he also was keenly aware of ecological issues and integrated genetics with ecology in the 1960-1970s; he wrote the major textbook *Principles of Plant Breeding,* and served as president of several scientific societies, including the Genetics Society of America, the American Genetic Association, and the American Society of Naturalists.

AMOS, H. R. >>> BACKHOUSE, W. O.

ANDERSON, R. G. (1924-1981): born in Ontario (Canada); he worked on genetics of rust resistance in bread wheat at the Canada Department of Agriculture; in 1964, he was appointed by Rockefeller Foundation to work in India with the skilled assistance of Drs. M. S. SWAMINATHAN and S. P. KOHLI; he contributed to the concerns of politicians, wheat scientists, and farmers regarding feeding a hungry world.

B

BACKHOUSE, W. O. (1779-1844): an English geneticist and breeder who studied at Cambridge (United Kingdom) under J. BIFFEN; he moved to Argentina along with two assistants, J. WILLIAMSON and H. R. AMOS; their work was extremely valuable to Argentina, as genetic research was incorporated into breeding and new direction was given to agricultural experimentation; BACKHOUSE made extensive surveys throughout the wheat-growing regions, studying the characteristics of different populations and making single selection, mainly on "Barleta" and "Ruso" wheat populations; in 1925, he released a bread wheat cultivar "38 M. A."; it was the predominant cultivar in the wheat production of Argentina until 1944 and also a remarkable progenitor for a great number of new cultivars; the cultivar "Lin Calel M. A.", suitable for early sowing and grazing, was released in 1927; this cultivar was extensively grown in the center, south, and west of the province of Buenos Aires as well as in the province of La Pampa; its peak of popularity was in 1935 (21 percent of the total wheat production).

BAILEY, L. H. (1858-1954): born in South Haven, Michigan (United States), he was the youngest child on a fruit farm; in 1877, BAILEY left South Haven and began his secondary education at the Michigan Agricultural College (now Michigan State University); while at Michigan, he worked with W. J. BEAL, a botanist with whom he had become acquainted through participation in the Michigan State Pomological Society; he became interested in plant breeding and was involved in the Natural History Society; his first articles on identification of local flora were published in *The Botanical Gazette;* he began a long-term involvement with classification of the genus *Rubus;* after leaving Michigan Agricultural College, BAILEY went to Harvard University to work for the botanist A. GRAY, where he was responsible for sorting and classifying plant specimens received from Kew Gardens (England); by 1888, the dynamic and prolific BAILEY had a well-established reputation as a scientist and an innovator in the area of horticultural research and education; consequently, he was recruited to fill a new position as professor of horticulture and botany at Cornell University and as dean of the Agriculture College; he was one of the most influential botanists and horticulturists of the 20th century, having written countless authoritative articles and books; his *Manual of Cultivated Plants, Cyclopedia of Horticulture,* and *Hortus I-III* series being widely known; the term "cultivar" was coined by him.

BAKER, R. F. (1906-1999): the first maize breeder with PIONEER HI-BRED, one of the preeminent plant breeders of the 20th century, and a graduate of the Iowa State University, Agronomy Department (United States).

BARABAS, Z. (1926-1993): a Hungarian agronomist and plant breeder born in Budapest; after receiving his PhD from the University of Agricultural Sciences, he started his career there as a teaching assistant; from 1951, he began flax and sorghum breeding; in 1960, he became the head of the Hungarian Sorghum Breeding Program in Martonvasar and remained so for nine years during this time, he released four hybrid varieties; in 1969, he moved to Szeged and became the leader of the Wheat Breeding Program at the Cereal Research Institute of the Hungarian Ministry of Agriculture; he made important contributions to the development of eighteen wheat varieties; he also began a breeding program for durum wheat in Hungary; he had a wide range of interests, such as studies on genetic and physiological background of male sterility, hybrid vigor, as well as mutations in sorghum and wheat; he was the first to introduce genetically determined male sterility in a series of sorghum strains; as a breeder and director, he successfully established the periodical *Cereal Research Communications* (Szeged, Hungary) and edited it for more than twenty years.

BATESON, W. (1861-1926): born in Whitby (England) and educated at St. John's College, Cambridge; as the first English geneticist, he recognized the importance of MENDEL's pioneering work; he translated and introduced MENDEL's papers to Britain and coined the word "genetics" in a paper of 1907; he was the founder of the "Cambridge School" of genetics; in 1899, the Royal Horticultural Society organized an international conference on hybridization; referring to MENDEL's (thirty years before), BATESON stated in his paper: "What we first require is to know what happens when a variety is crossed with its nearest allies. If the result is to have a scientific value, it is necessary that the offspring of such crossing should then be examined statistically. It must be recorded how many of the offspring resembled each parent, and how many showed characters intermediate between those of the parents. If the parents differ in several characters, the offspring must be examined statistically, and marshaled, as it is called, in respect to each of those characters separately . . ." (In Hybridization and Cross-Breeding as a Method of Scientific Investigation. *J. Roy. Hort. Sco.* 24, p. 63); one year later, he introduced the term "allelomorphe" for the opposite characters of a pair of characters, the term "heterozygote" for nuclei that resemble the opposite "allelomorphes" (later abbreviated to "allele"), and the term "homozygote" for identical "allelomorphes;" he became the first director of the John Innes Horticultural Institute in Merton (England).

BAUR, E. (1875-1933): a German botanist who contributed much to the utilization of genetic knowledge in breeding; in 1894, he began to study medicine at the universities of Heidelberg, Freiburg, and Strasbourg; in 1897, he continued medical studies at the University of Kiel; at the same time, he attended lectures in botany and biology; in 1900, he received his MD from the University of Kiel; in 1903, he earned a doctorate with a study of the developmental aspects of fructification in lichens; in 1911, he received a professorship in botany at the "Landwirtschaftliche Hochschule" in Berlin, and in 1928, he became the director of the new Kaiser Wilhelm Institut für Züchtungsforschung at Müncheberg; from 1931, he held lectures on evolution, applied genetics, and eugenics in London, Sweden, and South America; he was the editor of *Zeitschrift für induktive Abstammungs- und Vererbungslehre* (from 1908 onward; now, Molecular and General Genetics) and founder of *Der Züchter* (1929, now, *Theoretical and Applied Genetics*); he was cofounder of the Deutsche Gesellschaft für Vererbungswissenschaft (German Society of Heredity) with C. CORRENS and R. GOLDSCHMIDT, and was elected as president of the 5th International Congress of Genetics (1927).

BEADLE, G. W. (1903-1989): he is credited as "the man who laid the foundation for the field of biotechnology"; he studied agronomy under F. KEIRN at the University of Nebraska, Lincoln (United States), and earned a BSc degree in 1926 and an MSc in 1927; his interest in genetics began at Cornell, where he began his graduate work and earned his PhD; in 1931, BEADLE was distinguished as a National Research Council Postdoctoral Fellow and instructed at California Institute of Technology until 1935; he served on the faculties of Harvard University, Stanford University, and the Institut de Biologie in Paris (France); during these years, he conducted his Nobel Prize–winning research [subjecting *Neurospora* (red bread mold) to X-rays in order to induce mutations; he then noted some mold ceased to produce specific organic compounds needed to survive; he added similar but different compounds and observed how the mold synthesized needed chemicals; his conclusion was that the characteristic function of the gene was to control the synthesis of a particular enzyme]; in 1941, he recognized the genetic control of biochemical reactions in *Neurospora;* in 1946, he returned to Caltech as professor and chair of the Division of Biology while he was serving as the George Eastman Professor at Oxford University; BEADLE received the 1958 Nobel Prize in Physiology and Medicine along with J. LEDERBERG and E. TATUM for their discovery that genes control chemical reactions by their formation of specific enzymes; thirty years before he was awarded the Nobel Prize, BEADLE was working with teosinte, a wild grass that is closely related to maize; this early work on teosinte began a lifelong fascination with the origin of maize and set him on a mission to confirm a hypothesis that he had settled in his own mind as a graduate student, that teosinte is the progenitor of cultivated maize; although the path of events led him away from this mission during the most productive years of his career, upon retirement from the presidency of the University of Chicago in 1968, he again took up research on teosinte, employing experimental genetics and organizing an expedition to Mexico in search of naturally occurring mutants in teosinte populations that might shed some light on the steps that transformed teosinte into maize.

BEAL, W. J. >>> BAILEY, L. H.

BEKE, F. (1914-1988): born in Bana (Hungary), he persued his studies at Gyoer and later on at the Agricultural High School at Mosonmagyarova; from 1949, he continued his research work at the Plant Breeding Institute, Fertoed, on clover, rape, and wheat; one of his best results was the improvement of the winter wheat variety "F293;" this variety with its excellent leaf rust resistance has been a leading variety in Hungary over almost twenty-nine years; the investigation of the relationships among plant development, growth, and productivity was his main concern.

BERZSENYI-JANOSITS, L. (1903-1982): an outstanding personality and father figure of Hungarian plant breeding; he was mainly concerned with Hungarian maize breeding and organization of the Research Institute Marosvasarhely for more than twenty years; after 1945, he reorganized Hungarian plant breeding; with the help of the Food and Agriculture Organization (FAO) of the United Nations, he provided the first foreign (American) inbred lines to Hungary.

BESSEY, C. A. >>> EMERSON, R. A.

BIFFEN, R. H. (1874-1949): he initiated the wheat breeding at the Plant Breeding Institute, Cambridge (United Kingdom), and was its first director in 1896; he introduced Mendelian genetics into wheat breeding; in 1905, he was the first to breed crops for rust resistance in cereals [the variability in the pathogen was not fully appreciated until the work of STAKMAN and PIEMEISAL in 1917, who reported that stem rust in cereals and grasses comprised six biological forms; these forms were distinguished from each other morphologically and parasitically and were differentiated on selected cereal and grass hosts; this led to research on variation and variability of plant pathogens and the breeding of plants for resistance to specific races]; BIFFEN bred famous wheat varieties including "Little Joss" (released in 1908) from a cross of the old English wheat "Squarehead's Master" and a Russian variety "Ghirka" with good resistance to yellow rust.

BLAKESLEE, A. F. >>> MÜNTZING, A.

BOHR, N. >>> DELBRÜCK, M.

BOHUTINSKY, G. (1877-1914): he was head of the Plant Breeding Station, and professor at the Royal Farming School at Krizevci (Croatia) in the period 1903-1912; he conducted extensive research work mainly on wheat and maize; his paper "The Crossing of Squarehead × Banatska Brkilja" (1911), published in *Gospodarska Smorta*, represents one of the earliest works on wheat genetics in the territory of former Yugoslavia; by the method of individual selection from autochthonous populations like "White Wheat from Srijem," "Somogy Wheat," or "Sirban Prolific," he created outstanding new varieties for Croatia, Istria, Dalmatia, Slovenia, and Bosnia; they were grown under the name of "Bohutinsky's Wheats" until 1925.

BOJANOWSKI, J. S. (1921-2004): he was born in Warsaw (Poland) and graduated from the Agricultural University in Warsaw in 1949 (MSc in Agronomy) with specialization in genetics and plant breeding; he received his PhD from the Agricultural Academy (Poznan) in 1963; he started breeding work in a private company in 1943, and continued until 1953 as its chief breeder; during this period he was working with several crops, such as cereals, fodder beets, fodder carrots, forage grasses, and alfalfa; in 1953, he joined the Department of Plant Genetics of the Polish Academy of Sciences, led by Prof. E. MALINOWSKI; from this time on, he concentrated his research on heterosis and hybrid maize breeding; in 1959, he moved to the Department of Genetics of the Agricultural University in Warsaw, where he remained until 1969 [because of the generally low agricultural productivity and political acceptance of genetic principles, the years 1953-1969 were very important for hybrid maize breeding in Poland; by the end of the decade, first Polish hybrids were registered]; during this period, BOJANOWSKI passed a one-year training at the Iowa State College in Ames (United States) under the leadership of Profs. G. F. SPRAGUE and W. A. RUSSELL; in 1969, he was appointed as deputy director of the Institute of Plant Breeding and Acclimatization (IHAR, which is the abbreviation of the Polish name) in Radzików, which played an important role in the maize-breeding program of Poland; in 1976, as the head of the Maize Department of IHAR, he initiated work on breeding of high-lysine maize, which continues to the present; BOJANOWSKI was a member of the Board of the European Association for Research on Plant Breeding (EUCARPIA) between 1975 and 1981, and the president of EUCARPIA between 1983 and 1986; he developed seven varieties of agricultural crops: one of spring barley, one of oats, and five of maize; he translated to Polish two important textbooks on plant breeding: F. C. ELLIOTT's *Plant Breeding and Cytogenetics* and R. W. ALLARD's *Principles of Plant Breeding.*

BOLIN, P. >>> NILSSON-EHLE, H.

BOOM, B. K. (1903-1980): a horticultural taxonomist at Wageningen (the Netherlands); a conifer expert and co-author of the landmark *Manual of Cultivated Conifers;* he did much to stabilize garden conifer nomenclature.

BORLAUG, N. E. (1914-): an American agricultural scientist born in Cresco, Iowa (United States), he was a plant pathologist and the winner of the Nobel Prize for Peace in 1970; he studied plant biology and forestry at the University of Minnesota and earned a PhD in plant pathology (1941); in 1943, BORLAUG joined the staff of the Cooperative Mexican Agricultural Program as an employee of the

Rockefeller Foundation; this program resulted from a trip to Mexico by U.S. Vice President H. A. WALLACE and a request from Mexico for technical assistance to improve its agricultural research; in 1944, he arrived in Mexico; BORLAUG labored for thirteen years before he and his team of agricultural scientists developed a disease-resistant wheat [native wheats produced low yields, and in an attempt to solve this problem, BORLAUG turned to several Japanese dwarf strains, which he crossed with his rust-resistant variety; the result was a hard spring wheat that resisted rust, tolerated the climatic and soil variations across Mexico, and reduced toppling; it produced large yields with the use of nitrogen fertilizer and irrigation]; as a result, in 1956, Mexico achieved self-sufficiency in wheat production; he also created new wheat–rye hybrids for cereal production known as triticale; he served as director of the Inter-American Food Crop Program (1960-1963) and as a director of the International Maize and Wheat Improvement Center at Mexico City, Mexico (1964-1979); he was one of those who laid the ground work of the so-called "Green Revolution," the agricultural technological advance that promised to alleviate world hunger.

BORODIN, I. P. >>> NAVASHIN, S. G.

BOTHA, L. >>> NEETHLING, J. H.

BOVERI, T. >>> SUTTON, W.

BRIDGES, C. B. >>> RHOADES, M. M.

BRUUN von NEERGAARD, T. >>> NILSSON-EHLE, H.

BURBANK, L. (1849-1926): an American plant breeder who developed over 800 new strains of plants, including many popular varieties of potato, plums, prunes, berries, as well as trees and flowers; one of his greatest inventions was the "Russet Burbank" potato (also called the "Idaho potato"), which he developed in 1871; this blight-resistant potato helped Ireland recover from its devastating potato famine of 1840-1860; he also developed the "Flaming Gold" nectarine, the "Santa Rosa" plum, and the "Shasta" daisy; being raised on a farm and only attending elementary school, he was self-educated; he applied Mendelian genetics to his breeding approaches; on DARWIN's *The Variation of Animals and Plants under Domestication,* BURBANK said, "It opened up a new world to me."

BUSTARRET, J. (1904-1988): born in Bordeaux (France); he enjoyed a school career at the Lycée Montaigne; poor health led him the direction of the Institute National Agronomique, where he was admitted in 1924 and graduated in 1926; in 1930, he joined the plant-breeding station at Epoisses/Dijon; in 1938, BUSTARRET joined the Agricultural Research Center at Versailles [his team worked on resistance to cold and bunt in wheat and resistance to smut in oats; it also provided operational instructions for assessing varieties and on the recommended methods and techniques for breeding]; from 1950, his wheat variety "Etoile de Choisy" played an important role in the renewal of agriculture in southwestern France; from 1937, BUSTARRET also carried out resistance breeding to viruses and blight in the potato; he created an excellent variety that is still highly appreciated today—"BF 15"; one of BUSTARRET's great concerns (when he was in charge of the Central Plant Breeding Station from 1944) was the training of young scientists at the National School of Horticulture (Ecole Nationale Supérieure d'Horticulture—ENSH), which he accomplised from 1941 to 1959; he was one of those who inspired the policy of the Standing Technical Committee on Plant Breeding (Comité Technique Permanent de la Sélection, CTPS) from 1942 [it helped standardize breeding and propagation methods in France; it was behind the creation of a National Field Experiment Network since 1950]; the International Convention for the Protection of New Varieties of Plants (the so-called "Paris Convention," adopted in 1961) contributed to the inauguration of European Association for Research on Plant Breeding (EUCARPIA), participation in which crowned BUSTARRET's international activity [it was created on the initiative of BUSTARRET, C. DORST (Wageningen, the Netherlands), and of >>> W. RUDORF (Köln, Germany)]; from 1961 to 1964, BUSTARRET was chairman of this association, which today the most important body, well beyond the frontiers of Europe, for worldwide concertation in the plant-breeding sector.

CAMARA, A. >>> MELLO-SAMPAYO, T.

CARRIERE, E. A. (1818-1896): a French conifer expert, author of the well-known *Traité Générale des Conifères* and of numerous Latin cultivar names still in widespread use today; he was one of the most important scholars of cultivated varieties of his time.

CATCHESIDE, D. G. (1907-1994): he was one of the seminal persons in the postwar development of genetics, both in the United Kingdom and Australia; he made distinguished contributions to plant genetics and cytology, to genetic effects of radiation, fungal biochemical genetics, control of genetic recombination, and, in his retirement, to bryology; as a professor and administrator, he was responsible for several new institutional developments including the first Australian Department of Genetics, the first Department of Microbiology at Birmingham (United Kingdom) and, perhaps most importantly, the Research School of Biological Sciences of the Australian National University.

CORRENS, C. E. (1864-1934): a German botanist and geneticist; in 1900, independent of but simultaneously with the biologists TSCHERMAK and >>> H. de VRIES, rediscovered MENDEL's historic paper on the principles of heredity; he was born in München and raised by his aunt in Switzerland; in 1885, he entered the University of München to study botany; C. NÄGELI, the botanist to whom MENDEL wrote to about his pea plant experiments, was no longer lecturing at München; NÄGELI, however, knew CORRENS' parents and took an interest in him; NÄGELI was the one who encouraged CORRENS' interest in botany and advised him on his thesis subject; later CORRENS became a tutor at the University of Tübingen (Germany), where he began to deal with trait inheritance in plants in 1893; he already knew about some of MENDEL's hawkweed plant experiments from NÄGELI; NÄGELI never talked about MENDEL's key pea plant results, so he was initially unaware of MENDEL's laws of heredity; it changed by 1900, when CORRENS submitted his own results for publication, the paper was called "Gregor Mendels Regel über das Verhalten der Nachkommenschaft der Bastarde" (Gregor Mendel's Law Concerning the Behavior of the Progeny of Hybrids);

CORRENS was active in genetic research in Germany, and was modest enough to never have a problem with scientific credit or recognition; he believed that his other work was more important, and the rediscovery of MENDEL's laws only helped him with his other work; he was supposedly indignant that H. de VRIES did not mention G. MENDEL in his first printing; in 1913, CORRENS became the first director of the newly founded Kaiser Wilhelm Institut für Biologie in Berlin/Dahlem (Germany); unfortunately, most of his work was unpublished and destroyed when Berlin was bombed in 1945.

COTTA, H. (1763-1844): a German forester; founder of scientific and practical forestry in Europe at the Tharandt School of Forestry (Germany).

CRICK, F. H. C. (1916-2004): an English scientist, born in Northampton; he attended University College in London and received a BSc in physics (1937); in 1947, he began to study biology; he joined the Cavendish Laboratory (Cambridge) in 1949, where he started work on his PhD thesis studying proteins by X-ray diffraction; it was at the Cavendish Laboratory where he and >>> J. WATSON met; they soon realized their common interest in the nature of the "genetic material;" there was still a debate among scientists as to whether proteins or DNA were the genetic material; even without knowing the identity of the molecule, CRICK and WATSON were among those who thought nucleic acids were the key to hereditary transfer; with the experiments in 1952, they, and others, became more sure of this hunch; they wanted to determine the structure of DNA and hoped this would lead to a deeper understanding of how life is propagated.

CURTIS, W. (1746-1799): founder of *Botanical Magazine* (United States) in 1787, one of the first periodicals devoted to ornamental and exotic plants.

DARLINGTON, C. D. (1903-1981): an English cytogeneticist born in Chorley; educated at Mercer's School, Holborn, 1912-1917, St. Paul's School, 1917-1920, and Wye College, Ashford, 1920-1923; in 1923, he began an association of more than thirty years with the John Innes Horticultural Institution; in 1937, he became head of the

Cytology Department and Director of the Institute in 1939; much of his work on cytology and chromosome theory was augmented by expeditions and work abroad; in 1953, he resigned from the institution and accepted the Sherardian Professorship of Botany at Oxford; in addition to his research, teaching, and publication, he took a keen interest in the Botanic Garden and created the "Genetic Garden."

DARWIN, C. (1809-1882): an English biologist who wrote the book *The Origin of Species* in 1859; he first presented the idea in a scientifically plausible way that biological species can change over the time.

DELBRÜCK, M. (1906-1981): a German biologist born in Berlin; his father was H. DELBRÜCK, a professor of history at the University of Berlin, and his mother was the granddaughter of Justus von LIEBIG, a famous German chemist and plant physiologist; DELBRÜCK studied astrophysics, shifting towards theoretical physics, at the University of Göttingen; after receiving his PhD, he traveled through England, Denmark, and Switzerland, where he met W. PAULI and N. BOHR, who got him interested in biology; DELBRÜCK went back to Berlin in 1932 as an assistant to L. MEITNER; in 1937, he moved to the United States, taking up research at Caltech on *Drosophila* genetics; in 1942, he and S. LURIA demonstrated that bacterial resistance to viral infection is caused by random mutation and not adaptive change—for that, they were awarded the Nobel Prize in Physiology and Medicine in 1969; from the 1950s, DELBRÜCK worked on physiology rather than genetics; he also set up the Institute for Molecular Genetics at the University of Köln (Germany).

DICKSON, A. D. (1900-1997): he was the first director of the Agricultural Research Service, Barley and Malt Laboratory at the United States Department of Agriculture (USDA), serving from its inception in 1948 until his retirement in 1968; he was born in Moxee City, Washington; he earned his BSc in 1919, and in 1929, his PhD in biochemistry from the University of Wisconsin, Madison; after a brief period in Washington, he returned to Wisconsin in 1931, where he took a position with the USDA; together with his colleagues, he initiated a barley testing and research program directed towards two goals: (1) developing procedures and equipment for evaluating the malting quality on a laboratory scale and (2) comparing the new hybrids with the older varieties "Oderbrucher" and "Manchuria"; he conducted research on physical and chemical variables of malt, particularly the amylases, as influenced by variations in the malting process; his major accomplishments were the establishment of the Agricultural Research Service (ARS) barley-quality-testing program and the development of a close relationship with the barley breeders and the malting and brewing industry; he also did research on *Fusarium* toxins in scab-infested barley.

DIPPE, A. (1824-1890): a German agronomist and breeder who founded a breeding and seed company in Quedlinburg (Germany); it was one of the biggest seed producers for sugar beets between 1870 and 1920.

DIPPE, F. C. von (1855-1934): a German horticulturist at Quedlinburg (Germany); after a gardener apprenticeship in Stendal, he attended to the family enterprise "Gebrüder Dippe," where he became manager of agriculture and the breeding of flowers and vegetables; in the 1880s he became manager of the crop-breeding program; in 1890, he began work with individual plant selection and examination of the progeny.

DOBZHANSKY, T. >>> RHOADES, M. M.

DOROFEEV, V. F. (1919-1987): a Russian agronomist, botanist and expert in wheat breeding; he was the head of the Department of Wheat (1978-1987) and director of N. I. Vavilov All-Union Research Institute of Plant Industry (VIR) (Leningrad, Soviet Union).

DUBININ, N. P. >>> KHVOSTOVA, V.

EAST, E. M. (1879-1938): born in Du Quoin, Illinois (United States); from 1897 he attended the Case School of Applied Science in Cleveland, and from 1898, he studied chemistry at the University of Illinois (BSc 1900, MSc 1904, PhD 1907); from 1905 to 1909, he was an assistant at the Connecticut Agricultural Experiment Station, and moved to Harvard University in 1914, where he became a professor of biochemistry; most known for his book *The Role of Selection in Plant Breeding*.

EIHFELD, J. G. (1893-1989): a Russian biologist and plant breeder; from 1923 to 1940, he was in charge of the Polar Branch of VIR (Leningrad, Soviet Union), and from 1940 to 1951, director of VIR headquarters.

ELLIOTT, F. C. >>> BOJANOWSKI, J. S.

ELLISON, F. (1941-2002): he had a long association with the University of Sydney Plant Breeding Institute (Australia); he received his BSc in Agriculture in 1967, his MSc in 1971, and PhD in 1977; in 1975, he was appointed as Assistant Wheat Breeder and undertook postdoctoral research in wheat breeding at the University of Manitoba, Winnipeg (Canada); in 1976, he was appointed as plant breeder, and he was promoted to Senior Wheat Breeder in 1988; his efforts were particularly directed at the release of prime hard wheats for northern New South Wales and Queensland (Australia); he bred the wheat cultivars "Sunkota," "Shortim," "Suneca," "Sunstar," "Sundor," "Sunbird," "Sunelg," "Sunco," "Sunfield," "Miskle," "Sunbri," "Sunmist," "Sunstate," "Sunland," "Sunvale," "Sunbrook," "Sunlin," "Sunsoft 98," "Braewood," and "Marombi"; ELLISON also assisted the efforts of N. DARVEY and R. JESSOP (University of New England, Armidale) in the development of triticale cultivars "Ningadhu," "Samson," "Bejon," "Madonna," and "Maiden."

EMERSON, R. A. (1873-1947): born at Pillar Point, New York (United States), but it was Nebraska that nurtured his early development and schooling; he spent fifteen years of his professional career at the University of Nebraska, followed by thirty-three years at Cornell University; in 1893, he enrolled in the College of Agriculture at the University of Nebraska; he came under the influence of C. A. BESSEY; in 1897, he received his BSc; his first job was in the Office of Experiment Stations (USDA, Washington, DC); he served there for two years (1897-1899) as assistant editor of *Horticulture;* in 1899, he accepted an appointment at the University of Nebraska as assistant professor and chairman of the Horticulture Department, and horticulturist in the Experiment Station; he held these positions for fifteen years; he covered a wide range of horticultural projects, such as different culture methods for fruits and vegetables, domesticating native wild fruits, or winter hardiness of trees; he began to hybridize garden beans in 1898, while he was in Washington, DC, to find out if there were any definite principles controlling heredity in plants; in 1902, he first became aware of MENDEL's work because he referred to it in a paper published in the 15th Annual Report of Nebraska Experiment Station; he quickly realized that his own studies on beans could be used to test MENDEL's principles and confirmed MENDEL's observation that some characters were dominant over the alternative forms; in 1909, EMERSON summarized his findings on "Inheritance of color in the seeds of the common bean, *Phaseolus vulgaris*" in the Annual Report of Nebraska Experiment Station; EMERSON's interest in maize for studies of heredity began around 1908, when he grew some plants from a cross between a rice popcorn and a sweet corn variety for a teaching demonstration; the segregation of starchy and sugary kernels deviated from the expected ratio based on a single-factor pair; he collected a wide variety of genetic deviants in maize, many of which were used in later studies at Cornell University; at the time, when the relationship between genetic linkage and chromosomes had not jelled, he presented a report and stated that "if genes were definitely located in chromosomes and that if parental chromosomes separated bodily at the reduction division, we should have an explanation not only of perfect genetic correlation and of allelomorphism but of independent inheritance as well" (In Inheritance of a Recurring Somatic Variation in Variegated Pericarp of Maize. *Amer. Nat.* 48, 1914, p. 112); in 1908, by making some crosses between a Missouri dent corn and two dwarf types of popcorn, he initiated a study on quantitative inheritance in maize; in 1911, he spent a year at Bussey Institution of Harvard University pursuing a graduate program for a DSc (awarded in 1913); his advisor was the distinguished geneticist, >>> Prof. E. M. EAST, whose special interest was quantitative inheritance; EAST, from studies on endosperm color in maize, and the Swedish geneticist >>> NILSSON-EHLE, from studies on color segregations in wheat and oats, independently proposed what came to be known as the "multiple-factor hypothesis" to explain the inheritance of quantitative characters; they assumed that the continuous variation in segregating progenies was governed by several to many genes that were cumulative in their action; EMERSON and EAST collaborated in compiling *Nebraska Research Bulletin No. 2* on "The inheritance of quantitative characters in maize," which was published in 1913; because of his merits, in 1914 he accepted an offer from Cornell University at Ithaca, New York (United States) to become professor and head of the Department of Plant Breeding, positions that he was to hold until his retirement in 1942; he recommended a number of proposals to develop the Plant Breeding Department, such as inaugurating clonal selection with fruit and flower crops, scion selection with apples, making biochemical investigations of color inheritance, and developing a botanical garden to illustrate MENDEL's experiments; in 1924, he and F. D. RICHEY, USDA maize investigator, made a scientific expedition to Argentina, Bolivia, Chile, and Peru; they collected around 200 samples of maize; in 1935, *A Summary of Linkage Studies in Maize* was compiled by EMERSON, G. W. BEADLE, and A. C. FRASER; for the first time, all the known genes (>300) in maize were catalogued alphabetically with appropriate symbols, descriptions, and chromosomal locations when known; EMERSON was cited for forty-one of these genes, which was indicative of the many genetic studies on mutant types that he had made through the years; for a few years (1938-1942) he was associated with a project on breeding muskmelons for resistance to *Fusarium* wilt.

ENGLEDOW, F. L. >>> HUNTER, H.

EYAL, Z. (1937-1999): a prominent leader of plant pathological research in cereals and its practical applications in combating fungal disease; he studied agronomy at Oklahoma State University, Stillwater (United States) and earned a PhD in plant pathology in 1966; he served twice as a head of Department of Plant Sciences of Tel Aviv University (Israel) and was appointed as director of the University's Institute for Cereal Crops Improvement in 1996; he became one of the leading researchers in the field of wheat–*Septoria* interactions, initiating and guiding approaches toward understanding both plant and pathogen biology.

FENZL, E. >>> TSCHERMAK-SEYSENEGG, E. von

FIALA, J. L. (1924-1990): a Catholic priest from the United States and one of the century's leading hybridizers of *Malus* and *Syringa;* he wrote remarkable books on both genera; his lilac-breeding program spanned more than fifty years and produced great results; "Avalanche," for example, is one of the top-rated white lilacs.

FIFE, D. (1805-1877): a Scottish farmer and wheat breeder; he moved to Canada with his parents in 1820 and settled in Otonabee/ Peterbourgh (Ontario); FIFE brought Russian spring wheat to Canada; he grew it on his farm and harvested half a bushel of grain, which he shared, in part, with his friends; in 1842, this wheat came to Russia from Ukraine; the original strain was called "Halychanka" and later known as "Red Fife" [records on the "Halychanka" variety were found in the Galician chronicle as far back as the time of King Yaroslav OSMOMYSL (1171-1187); the fact that the genetic characteristics of the "Halychanka" variety are based on a selection process dating back to the twelfth century shows why this variety is so unique; its stability (genetic homozygosity) deserved the attention of every plant breeder; the variety was cultivated over centuries in the fields of Ukrainian peasants; the strain was used widely after 1848, becoming the leader in Ontario by 1851 and virtually replacing all others there by 1860]; from FIFE's farm in Otonabee it spread to Illinois and Ohio in the United States, and then to Saskatchewan, Alberta, and Manitoba in Canada about 1870, ranking as the leading variety there from 1882 to 1909; "Red Fife" also served as the male parent of the "Marquis" strains, which proved more frost tolerant and even less susceptible to rusts, allowing wheat farming in Manitoba to spread further west and north.

FISHER, R. A. (1890-1962): he was born at St. James, London; in 1904, he entered Harrow and won the Neeld Medal in 1906 in a mathematical essay competition; FISHER was awarded a scholarship from Caius and Gonville College, Cambridge, which was necessary to finance his studies; in 1909, he matriculated at Cambridge; although he studied mathematics and astronomy, he was interested in biology; in his second year as an undergraduate, he began consulting senior members of the university about the possibility of forming a Cambridge University Eugenics Society; he passed with distinction the mathematical tripos of 1912; it was FISHER's interest in the theory of errors that eventually led him to investigate statistical problems; after leaving Cambridge, FISHER had no means of financial support, so he worked for a few months on a farm in Canada; later he returned to London, taking up a position as a statistician in the Mercantile and General Investment Company; the interest in eugenics and his experiences working on the Canadian farm made him start a farm of his own and give up being a mathematics teacher in 1919; at that time, two posts simultaneously were offered to him—the first by K. PEARSON as chief statistician at the Galton laboratories and the second as statistician at the Rothamsted Agricultural Experiment Station; the latter was the oldest agricultural research institute in the United Kingdom (see Table 3.1), studing the effects of nutrition and soil types on plant fertility, and it appealed to FISHER's interest in farming; he accepted the post at Rothamsted, where he made many contributions both to statistics (in particular to the design and analysis of experiments) and to genetics; there he introduced the concept of randomization and the analysis of variance [FISHER's idea was to arrange an experiment as a set of partitioned subexperiments that differed from each other in having one or several factors or treatments applied to them; the subexperiments were designed in such a way as to permit differences in their outcome attributible to the different factors or combinations of factors by means of statistical analysis]; in 1921, he proposed the concept of likelihood [the likelihood of a parameter is proportional to the probability of the data, and it gives a function, which usually has a single maximum value, which he called the "maximum likelihood"]; one year later, he gave a new definition of statistics as a method to reduce data [he identified three fundamental problems: (a) specification of the kind of population that the data came from, (b) data estimation, and (c) data distribution]; FISHER published a number of important books, e.g., *Statistical Methods for Research Workers* (1925), *The Design of Experiments* (1935), and *Statistical Tables* (1947); during his time at Rothamsted he conducted breeding experiments with mice, snails, and poultry, and the results he obtained led to theories about gene dominance and fitness, which were published in *The Genetical Theory of Natural Selection* (1930); in 1933, K. PEARSON retired as professor of eugenics at University College and FISHER was appointed to the chair; he held this post for ten years, before being appointed as Arthur Balfour Professor of Genetics at the University of Cambridge in 1943; he retired from his Cambridge chair in 1957; he then moved to the University of

Adelaide (Australia), where he continued his research for the final three years of his life.

FORREST, G. (1873-1932): an American plant collector, explorer, and *Rhododendron* expert; he was perhaps the leading introducer of new *Rhododendron* taxa to horticulture; his collections are still studied today.

FORTUNE, R. (1812-1880): curator of the Chelsea Physic Garden (United States); a plant explorer to China beginning in 1852; most plants bearing the epithet *fortunei* are in his honor; he introduced *Camellia sinensis*, the tea plant, to the West.

FRANKEL, O. (1900-1998): born in Vienna (Austria), he studied agriculture at the University of Berlin (Germany) and earned his doctorate in agriculture there in 1925; he was employed for two years (1925-1927) as a plant breeder on a large private estate at Dioseg/ Bratislava (Slovakia); later he began wheat and barley breeding at Lincoln College/Christchurch (New Zealand), where he worked until 1951; he began his breeding program with introducing quantitative assessments of grain yield, milling, and baking quality, which led to the release of the widely grown varieties "Cross 7" (1934), "Taiaroa" and "Tainui" (1939), "Fife-Tuscan" (1941), and "WRI-Yielder" (1947); he put a considerable effort into optimizing the role of quality testing in the selection process; his postretirement research was focused on the base-sterile mutants of speltoid wheats and examined the photoperiodic effects on floral initiation and floret sterility; he was one of the pioneers of the genetic resources movement (the conservation of biological diversity); from 1951 to 1962, he was chief of the Commonwealth Scientific and Industrial Research Organisation (CSIRO) Division of Plant Industry (Australia) and on the CSIRO Executive board (1962-1966).

FRASER, A. C. >>> EMERSON, R. A.

FREISLEBEN, R. (1906-1943): a German botanist working at the Institute of Plant Production of the University Halle/Saale (Germany); he died early in the Second World War; during his short life as cytogeneticist, he used X-ray treatment to select (together with >>> A. LEIN) a mutant line of barley showing stable mildew resistance (*mlo*); this so-called *mlo* locus is still present in many advanced barley variety around the world.

FRIEDRICH, B. (1899-1980): he studied plant breeding during the 1920s; during his study he assisted Prof. TSCHERMAK at the "Hochschule für Bodenkunde" in Vienna (Austria); from 1938 to 1948, he was working as barley and wheat breeder at Sladkowicovo (Czech Republic); after 1945 he received an appointment in Martonvasar (Hungary), where he remained until his retirement; he was active in searching for new techniques in plant breeding, and in applying new ideas to the mechanization of nursery techniques; In 1946, he successfully planted rust nurseries of wheat in naturally infected locations for detection of resistant genotypes.

FRUWIRTH, C. (1862-1930): professor of plant production at the University of Vienna (Austria); in 1905, he founded the Königliche Württembergische Saatzuchtanstalt at Hohenheim (Germany)—an institution that still contributes to successful breeding in Germany (see Table 3.1); he returned to Austria and bought the estate Walddorf where he started cereal breeding; in addition, he received particular reputation as a scientific author and as founder of the *Handbuch der Züchtung Landwirtschaftlichen Kulturpflanzen* (P. PAREY Verl., Berlin), with five volumes and seven editions; with the same publisher, he founded the journal *Zeitschrift Pflanzenzüchtung* in 1913 (now, *Plant Breeding*).

G

GALLESIO, G. (1772-1839): an Italian naturalist and pomologist; made many hybridization experiments with carnations; he was born in Liguria and died in Florin; although he studied jurisprudence and was active in several political positions, in 1811 he published his first book, *Traite du Citrus,* and in 1816 a second book, *Teoria Della Riproduzione Vegetale,* in which he described his observations, e.g., ". . . I have crossed white with red flowering carnations and reciprocally . . . the seeds I produced brought carnations with mixed colors . . . the plants sometimes show the characters of the other parent, depending which one is dominating . . ."—showing that the term "dominate" was already in use before it was applied by M. SAGERET in 1826 and by T. A. KNIGHT.

GALTON, F. (1822-1911): a British geneticist and a cousin of C. DARWIN; his statistical analysis of genetic segregation patterns led him to the introduction of the "correlation coefficient;" he also coined the word "eugenics."

GOLDSCHMIDT, R. >>> BAUR, E.

GOULDEN, C. H. (1897-1981): a geneticist born at Bridgend, Wales (United Kingdom); he took the course for farmers at the University of Saskatoon (Canada) and went on to a PhD in plant breeding before becoming chief cereal breeder at the Dominion Rust Research Laboratory; in 1925, he succeeded L. H. NEWMAN as dominion cerealist; in 1948, he became assistant deputy minister for research in the Department of Agriculture; as a natural mathematician, he took up the new specialty of biostatistics and wrote the first North American textbook on this subject in 1937 for the students he taught at the University of Manitoba; as the head of cereal breeding at Winnipeg for twenty-three years, GOULDEN was responsible for the creation of "Renown," "Regent," and "Redman" wheats, suitable for the Canadian climate and possessing various rust-resistant qualities; he also developed six varieties of rust-resistant oats.

GOVOROV, L. I. (1885-1941): a Russian agronomist and plant breeder; from 1915, he worked at Moscow Breeding Station; from 1923, he was the head of the Steppe Experiment Station, and later he was in charge of the Department of Leguminous Crops at VIR (Leningrad, Soviet Union).

GRABNER, E. (1878-1955): considered the doyen of Hungarian plant breeding; he started the breeding work in Hungary at the beginning of the 20th century; influenced by visits to European plant breeding institutes in Sweden and Austria, he created the prerequisites for professional plant breeding in Hungary based on MENDEL's laws of inheritance [summarized in his book *Breeding of Agricultural Plants,* published in 1908]; soon after 1908, the first Hungarian Institute of Plant Breeding was founded in Magyarovar in 1909 (see Table 3.1); he directed this institute for twenty-eight years; owing to his inspiring work in twenty locations in the country, professional plant breeding was carried out on several crops in 1911; based on his activities, the first Hungarian order regulating the system of variety registration was passed in 1915.

GRAY, A. >>> BAILEY, L. H.

GREBENSCIKOV, I. S. (1912-1986): born in St. Petersburg (Russia); he studied at the Agricultural Faculty of the University of Belgrade (Serbia); in 1938, he obtained his MSc; in 1942 he continued his work at the Genetic Department of the Kaiser Wilhelm Institut, Berlin (Germany), where he carried out human brain research together with his Russian colleague N. TIMOFEEFF-RESSOVSKY (one of the early mutation researchers); in 1946, he was appointed by the Zentralinstitut für Kulturpflanzenforschung, Gatersleben (Germany); he became a recognized specialist for taxonomic studies in maize and *Cucurbitaceae;* his particular interest was focused on the inheritance of quantitative traits, e.g., the ontogenetic dominance variance; he was one of the first taxonomists to introduce the term "convariety" into maize taxonomy; both in maize and *Cucurbitaceae,* he studied the phenomenon of heterosis.

GREW, N. (1641-1712): cofounder of the discipline of Plant Anatomy with MALPIGHI; he was born in Coventry (United Kingdom); he was a practicing physician first in Coventry and then in London and became secretary of the Royal Society of London; his work on plant anatomy began in 1664 with the objective of comparing plant and animal tissues; an essay was published by the Royal Society of London in 1670 [MALPIGHI, working independently on the same subject in Italy, also sent his work to the Royal Society]; GREW approached botany from the medical standpoint [his fundamental thesis was that every plant organ consists of two "organical parts essentially distinct," i.e., a "pithy part" and a "ligneous part;" in the seed, the pithy part is composed of "parenchyma," a term first used by GREW]; he described stages of seed germination although the underlying physiology was hopelessly confused [he used the term "radicle" for embryonic root and "plume" for "plumule;" he called cotyledons "leaves" and recognized that they could appear above ground and turn green; he observed monocot stems with scattered (vascular) bundles and a lack of distinct bark and pith, resin ducts in cortex pine stem, wings, and "feathers" on seeds and fruit, protection and economy of space gained by overlapping of bud scales, folding and rolling of leaves in buds, and buds formed months before they expand (". . . a bulb is, as it were, a great bud under ground . . ."); he described a tulip flower in bulb in September and noted that pollen grains are "bee-bread;" he believed that micropyle allowed water to enter the seed and cause germination]; GREW initiated the study of tissues (histology); his major contribution was to relate anatomy and physiology; he made the first successful attempt to extract chlorophyll from leaves using oil as a solvent; his important work was *Anatomy of Plants* (1682).

GRUNDY, P. M. >>> YATES, F.

GYORFFY, B. (1911-1970): a Hungarian geneticist who obtained his PhD from the University of Szeged; in 1937, he obtained a post-doctoral position at the Kaiser Wilhelm Institut (Berlin, Germany); >>> F. von WETTSTEIN, a former student of C. CORRENS, strongly influenced GYORFFY's scientific career; GYORFFY was inspired by the numerous scientific papers about induced polyploidization via colchicine treatment; in 1944, he became director of the National Institute of Plant Breeding in Magyarovar (Hungary); in several positions he contributed very much to the development of the modern Hungarian plant breeding and agriculture.

HADJINOV, M. I. (1899-1988): an outstanding Russian maize breeder and geneticist, began working in 1940 at the Research Institute of Agriculture, Krasnodar (Soviet Union); from 1946 to 1948, he widely used inbred lines to obtain variety-line crosses that resulted in a number of commercial hybrids sown over large acreage in the Soviet Union; in 1954, he was attracted to studies and utilization of cytoplasmic male sterility (CMS) in maize breeding at almost the same time when in the United States the first publication on CMS for hybrid seed production appeared; he initiated studies on maize grain quality (opaque-2), polyploidy, distant hybridization with *Tripsacum* and *Teosinte*, induced mutagenesis, haploidy, etc.

HAKANSSON, A. >>> NILSSON-EHLE, H

HALDANE, J. B. S. (1892-1964): a native of Oxford (United Kingdom); he studied mathematics and biology at the University of Oxford; he carried out several genetic studies in plants; after his time as lecturer on enzyme research at Cambridge University, he became the successor of W. BATESON at the John Innes Horticultural Institute at Merton near London; from 1927 to 1936, he carried out intensive genetic research; in 1933, he became head of the Chair of Genetics (later, Chair of Biochemistry) at University College London; from 1957, he was an active supervisor at the Indian Federal Bureau, Calcutta, and founded the Laboratory of Genetics and Biometrics in Bhubaneswar (Orissa), where he died.

HALES, S. (1671-1761): an English physiologist, chemist, inventor, and country vicar; in 1709, he resigned a fellowship at Cambridge University to become a perpetual curate at Teddington; he studied physiology based on the foundation of >>> GREW's work on plant anatomy; for forty years he devoted his leisure time to research in botany and zoology; his memoirs were published in a collected form as the *Statical Essays* dealing with problems of plant and animal physiology; *Vegetable Statics* (1727) became the classic work in plant physiology, and HALES is considered as the founder of experimental plant physiology; the greater part of his work is a record of successive experiments [an attempt to stop *"bleeding"* in a badly pruned grape vine by means of a piece of bladder tied over the wound gave him the idea for constructing a manometer; he found that root pressure showed a daily periodicity and was affected by changes in temperature; he noticed that leaves gave off water and proceeded to measure the amount of transpiration and to compare it with the amount absorbed by the root; he studied variations in the quantity of water transpired over a twenty-four-hour period and demonstrated reduction in transpiration at night]; he had definite notions of the part that the leaves played in plant nutrition and he studied leaf structure; he also contended that "plants very probably draw through their leaves some part of their nourishment from the air" and that leaves also absorbed light; HALES had a scientific mind of the highest order and is ranked along with GREW and MALPIGHI as an outstanding leader in botany and physiology of the eighteenth century; he was the first to use quantitative results in botanical experiments, such as movement of sap, transpiration, and flow of nutrients by girdling.

HAMMARLAND, C. >>> NILSSON-EHLE, H.

HÄNSEL, H. (1918-): an Austrian breeder born in Vienna; he started his professional career after studying languages and agriculture at the University of Agricultural Sciences in Vienna; he received his PhD in plant breeding in 1948; after several postdoctoral studies in Cambridge (United Kingdom) and Wageningen (the Netherlands), he decided to work as a practical plant breeder in the Probstdorfer Saatzucht Co. (Austria), located in the cereal growing area east of Vienna; breeding of bread and durum wheat and of barley became his passion; he never lost contact with his former University; in 1954, he received a DSc degree as a lecturer and later as an external professor of Plant Breeding; for more than thirty-three years he gave lectures on mutation breeding, breeding methodology, and developmental physiology; more than fifty-five varieties were bred by him; in wheat breeding he was able to combine high yielding performance with superior baking qualities in varieties such as "Probstdorfer Extrem," "Perlo," and "Capo"; at times, some of his varieties covered 75 percent of the wheat production areas in Austria of about 225,000 ha; his spring barley "Adora" and "Viva" combined excellent yield stability with durable resistances; in durum wheat breeding his varieties are cultivated now in Italy, France, and Spain.

HAVENER, R. D. (1930-2005): he was one of the pioneers in the global agricultural research system, working for the world's rural poor for more than five decades; he led CIMMYT (Mexico) from 1978 to 1985 as the center's third director general, bringing it recognition as one of the leading international agricultural research organizations in the world [when he came to CIMMYT, N. BORLAUG was director of the wheat and >>> E. SPRAGUE the director of the maize program; during his leadership, CIMMYT expanded its regional presence and strengthened the economics program]; for fourteen years he worked as a senior agricultural program officer of the Ford Foundation; he served as interim director general at both the Centro Internacional de Agricultura Tropical, Cali (Colombia) (1994) and the International Rice Research Institute, Los Banos (Philippines) (1998), and was instrumental in the founding of the International Center of Agricultural Research in the Dry Areas, Aleppo (Syria) and the International Livestock Research Institute, Nairobi (Kenya); he served as chair of the ICARDA Board of Trustees from 1999 to 2003 and was the founding president of the Winrock International Institute for Agricultural Development, a fellow of the American Association for the Advancement of Science, an advisor for the World Food Prize and sat on the Board of Directors of Sasakawa Africa Association [whose president is N. BORLAUG].

HEINE, F. (1840-1920): he was born in Halberstadt (Germany); after school from 1850 to 1859, he began to study agronomy at Ahlsdorf; as owner of the estate Emersleben, he started plant breeding in 1869; in 1885, he bought additional land in Hadmersleben and moved his seed company there; he was the founder of systematic plant breeding in Germany; he started mass and single plant selection and progeny testing in cereals, legumes and root crops; his wheat varieties "Heines Squarehead" (1872), "Heines Teverson" (1893), "Heines Rivetts" (1896), "Heines Kolben" (1871), "Heines Bordeaux" (1891), "Heines Noe" (1900), and "Heines Japhet" (1903) were standards for a long time in Germany; he also bred about twelve other varieties of crop, including spring barley, oats, winter rye, pea, broad bean, and sugar beet; the breeding station Hadmersleben has remained a famous place of plant breeding.

HELLRIEGEL, H. (1831-1895): a German from Mausitz, he studied chemistry at Tharandt and founded the Agricultural Experiment Station Dahme, where he became director in 1873; in 1886, together with H. WILFAHRT, he demonstrated the assimilation of atmospheric nitrogen by bacteria in the root nodules of legumes.

HERBERT, W. (1778-1847): he was a contemporary of >>> T. A. KNIGHT and a dean of Manchester of the Anglican church in England; he dealt with the question of whether or not the fertility of interspecific hybrids is a measure for the distance between the species [he came to the conclusion that the fertility or sterility of hybrids are not an evidence for the taxonomic relationships, and there are no borders between varieties and species concerning hybridization; in his opinion the environmental conditions in which a certain species grew up is more important for its successful hybridization than the systematic (genetic) borders between species and/or races: ". . . the only thing certain is, that we are ignorant of the origin of races . . . that God has revealed nothing to us on the subject . . . and that we may amuse ourselves with speculating thereon . . . but we cannot obtain negative proof, that is, proof that two creatures or vegetables of the same family did not descend from one source. But can we prove the affirmative . . . and that is the use of hybridizing experiments, which I have invariably suggested . . . for if I can produce a fertile offspring between two plants that botanists have reckoned fundamentally distinct, I consider that I have shown them to be one kind, and indeed I am inclined to think that, if a well-formed and healthy offspring proceeds at all from their union, it would be rash, to hold them of distinct origin" (In On Hybridization Amongst Vegetable, *J. Horticult. Soc*, London, 2, 1847, pp. 81-107)]; he was convinced that hybridization would improve the value of horticultural and agricultural crops—many years later he read the papers of KÖLREUTER, which confirmed this.

HOFFMANN, W. (1910-1974): professor of genetics and breeding research at University of Berlin (Germany); he started his career as junior assistant at the Botanical Institute Heidelberg (1933-1934) and later as assistant at the Kaiser Wilhelm Institut für Züchtungsforschung, Müncheberg (1935-1936); from 1942 to 1946, he was the head of the Department of Genetics, Breeding and Agronomy at Schönberg (Moravia); from 1946 to 1949, he worked as scientific officer at the Institute of Plant Breeding Hohenthurm (Germany), and later as full professor, 1950-1958; in 1958, he moved to Berlin, where he remained as professor and director of the Chair of Genetics and Breeding Research of the Technical University (from 1972, Institute of Applied Genetics of the Free University of Berlin) until 1974; his main interest was the systematic combination breeding in barley by utilization of induced mutants; in hemp, he developed pale-stalked mutants with good photoperiodic adaptability (released in 1940); moreover, he developed several allopolyploids in cabbage and wheat; his synthetic tetraploid *Brassica campestris* ssp. *pekinensis* × *B. campestris* ssp. *oleiferea* hybrid received much attention as fodder and green manure crop; it was released in 1969.

HOLDEFLEISS, P. (1865-1940): a German breeder at Salzmünde (Germany); in 1894, he received his PhD from the University of Halle/Saale after an apprenticeship in agriculture at the same university under >>> Prof. J. KÜHN; after qualification in 1897, he became assistant and professor at the Agricultural Institute and, in 1931, dean of the agricultural faculty.

HOLTKAMP, H. (1904-1988): master gardener and founder of the Holtkamp Greenhouses Co., the largest grower of African Violets (*Saintpaulia* species) in the world; he is credited with a number of notable innovations, including multiflorescence, semper florescence, and nondropping flowers, which have contributed significantly to the worldwide popularity of African Violets; he is remembered as one of the most influential pioneers in the African Violet industry, creating many new varieties, each year, in both the United States and Germany.

HOOKE, R. (1634-1703): an English experimental physicist with wide interests in science; he was the first to state clearly that the motion of heavenly bodies must be regarded as a mathematical problem, and he approached in a remarkable manner the discovery of universal gravitation; HOOKE had a strong personality and temper and made virulent attacks on I. NEWTON and other scientists, claiming that their published work was due to him; he is remembered in biology for the discovery and naming of "cells" as the units of plant structure; with the aid of a microscope he examined a wide range of materials and substances including feathers, lice, fleas, and cork; his observations were published in *Micrographia* (1665); he first recognized that charcoal, cork, and plant tissues were ". . . all perforated and porous, much like a honeycomb . . ."; to these pores he gave the name "cells," although the cell walls were not considered a constituent part.

HUNTER, H. (1882-1959): a British agricultural scientist who graduated in 1903 from Leeds University; he was appointed officer in charge of the barley investigations being conducted by the Department of Agriculture and Technical Instruction in Ireland; there he developed the variety of barley "Spratt-Archer," which was for many years the most widely grown malting barley in Britain; in 1919, he was appointed the head of the Plant Breeding Division of the Ministry of Agriculture for Northern Ireland, and in 1923, he moved to Cambridge to join R. H. BIFFEN, T. B. WOOD, and F. L. ENGLE-

DOW in the Plant Breeding Institute of the University School of Agriculture; HUNTER became the director of this Plant Breeding Institute in 1936; after his retirement in 1946, he served as the president of the Council of the National Institute of Agricultural Botany for three years, 1951-1953.

INNES, J. (1829-1904): he was the founder of John Innes Horticultural Institute; it became later the John Innes Institute of Plant Science, Norwich, United Kingdom; he was a city businessman working in partnership with his brother James Innes in a company that owned large sugarcane plantations in Jamaica and imported rum to England.

IVANOV, I. V. (1915-1998): professor of Agronomy and a member of the Bulgarian Agricultural Academy of Sciences; he was born in Karnobat (Bulgaria); he received his MSc (agronomy) in 1946 from Sofia University and his PhD, dealing with wheat breeding and seed production techniques, in 1974 from the Agricultural Academy; he started work in 1946 as an agronomist at the Institute of Scientific and Applied Research in Karnobat; in 1948, he was appointed at the Agricultural Institute of Dobrich to assist Prof. T. SHARKOV, a wheat breeder; he returned to the Institute in Karnobat from 1951 to 1962, when two wheat-breeding centers were formed at Sadovo and Dobrich; he served as chairman of the wheat-breeding program at the Agricultural Experimental Station in Sadovo (from 1962 until his retirement in 1976); between 1966 and 1969, he was the deputy director of the Agricultural Experiment Station in Proslav near Plovdiv; he contributed greatly to the development and release of twelve Bulgarian wheat varieties; his bread wheat cultivars "Sadovo 1" and "Katya" were the second and first place finishers in the 1977 and 1984 international field testing organized by the Agronomy Department at the University of Nebraska, Lincoln (United States); those cultivars are still grown in Bulgaria, Greece, and Turkey as good and productive wheats possessing a good balance of agronomic traits.

IVANOV, N. R. (1902-1978): a Russian plant scientist and plant breeder; from 1926, he worked at the Institute of Applied Botany (Leningrad, Soviet Union); he was the Institute's Director during the siege of Leningrad in the Second World War; from 1967, he was the scientific secretary of the Commission on N. I. VAVILOV's Heritage under the Academy of Sciences of the Soviet Union.

J

JENKINS, M. T. >>> RHOADES, M. M.

JENNINGS, H. S. (1868-1947): he began his career as a teacher after graduating from high school in his hometown of Tonica, Illinois (United States); in 1889, he received a position as assistant professor of botany and horticulture at Texas A&M College; during the decade in which he was pursuing the study of protozoan behavior, MENDEL's laws were rediscovered; in a series of fundamental, exhaustive, pioneer papers on *Paramecium*, published between 1908 and 1913, he laid broad and deep foundations for all subsequent genetic work; he showed that heredity and its problems are essentially the same in microorganisms as in plants; he formulated the "pure-line theory" for vegetative reproduction shortly after JOHANNSEN's comparable theory for sexual reproduction; he also analyzed the phenomenon of assortative mating and pointed out its role in the isolation of races; he was struck by the continued production of hereditarily diverse clones at conjugation, even after many successive inbreedings, and he undertook to examine the matter mathematically [as a result, general formulae for the diverse systems of mating were published in a series of papers between 1912 and 1917]; he dealt with selection and mutation, multiple factors and multiple alleles, the demise of the unit factor and representative particle interpretation of Mendelism, the inheritance of acquired characters, the interaction of genes and environment in the determination of the phenotype, and the limitations of evolution by loss.

JENSEN, N. F. (1915-): born in Hazen (United States), he studied at Cornell University until 1939; in 1946, he returned to Cornell and became assistant professor in the small-grains breeding program; in a career of more than more thirty years, he marked major achievements in the breeding of spring barley, winter barley, winter wheat, spring oats, and winter oats; he viewed the breeding process as being composed of three stages [he first divided it into heterozygous and homozygous phases, where the heterozygous phase deals with everything up to individual line selection (F5 or F6) and the homozygous phase deals with all subsequent line evaluation and variety release; he further subdivided the heterozygous stage, separating the planning and hybridization from the handling of the early generation progenies].

JESENKO, F. (1875-1932): he was a research assistant of >>> E. von TSCHERMAK in Vienna (Austria); he successfully crossed different wheat varieties with, e.g., *Triticum dicoccoides, Secale cereale* and *S. montanum;* by crossing "Mold Squarehead" wheat and "Petkus Rye," he was one of the first to produce perennial wheat–rye hybrids; after its backcrossing to wheat, a fertile perennial wheat plant was obtained; the first report of this extensive work was presented during the Fourth International Genetic Conference, Paris (France), in 1911; in 1919, the Faculty of Agriculture and Forestry was established at Zagreb University (Croatia), and F. JESENKO was appointed as lecturer; from 1921, he was professor of botany at Ljubljana University (Slovenia); his later research on interspecific hybridization, particularly studies on F1 plants from crosses *Triticum aestivum* and *Aegilops geniculata,* were never published due to his untimely death.

JOHANNSEN, W. L. (1859-1927): a Danish biologist from Copenhagen who called the phenomenon of dominance and recessiveness "genes"; he studied at the University of Copenhagen, where he later became a professor of botany; he conducted numerous crossing experiments, particularly with *Phaseolus vulgaris;* he was able to isolate four "pure lines," leading him to the theory about populations and pure lines; after 1900, he strongly promoted genetics as a science.

JOHNSON, R. (1935-2002): a British plant pathologist and wheat rust expert; he did groundbreaking research on durable resistance at the Plant Breeding Institute in Cambridge, where he started his career in 1964; following a brief interlude with UNILEVER Co. after the privatization of the Plant Breeding Institute, he continued research at the John Innes Centre in Norwich until his retirement in 1995; until his death, he continued to serve the scientific community as senior editor of *Plant Pathology* and chairman of the U.K. Cereal Pathogen Virulence Survey.

JOHNSON, V. A. (1921-2001): an outstanding American wheat breeder; he was born in Newman Grove, Nebraska; he graduated from Albion High School in 1939; he earned his BSc (1948) and PhD (1952) from the University of Nebraska at Lincoln; he was employed by the USDA and with the University of Nebraska, Department of Agronomy, from 1952 to 1986; he was a professor of agronomy and coordinator of the USDA-ARS Hard Red Winter Wheat Research Program; with longtime colleague Dr. J. SCHMIDT, he was coleader of the internationally recognized Nebraska Wheat Research Team [in more than thirty years of active service, this team developed and released twenty-eight new varieties of hard red winter wheat; notable varieties released during their tenure included "Scout," "Centurk," and "Brule"; the variety "Scout," released in 1963, was grown on more than 3 million ha, making it the most extensively grown cultivar in the United States at the time; the team also carried out pioneering research on the enhancement of nutritional value of wheat, on selection for yield stability, and on the development of hybrid wheat]; SCHMIDT was the principal organizer of five international wheat conferences sponsored by the U.S. Agency for International Development.

JOHNSTON, R. P. (1939-2001): a plant breeder of the Department of Primary Industries, Queensland (Australia); he studied for the BSc degree at the University of Queensland, graduating in 1961, and was immediately employed by the Department of Primary Industries as a linseed breeder; in 1967, he initiated a barley variety testing program with the intention of identifying material suited to malting barley production in the South Queensland cropping region; his PhD thesis "Single Plant Selection for Yield in Barley" was conducted at the University of Adelaide and accepted in 1973; upon his return to Queensland, he developed and led a full-scale barley breeding program; the barley varieties released from the program under his guidance include "Grimmett," "Tallon," "Gilbert," and "Lindwall."

KEIRN, F. >>> BEADLE, G. W.

KAPPERT, H. (1890-1976): a German botanist; from 1914 to 1920, he was an assistant of C. CORRENS in Berlin (Germany); as a university professor, scientist, and author of numerous papers and books, he heavily contributed to the breeding research and plant breeding in Germany.

KARPECHENKO, G. D. (1899-1942): born in Velsk (Russia); he was one of the leading geneticists during the 1930s, and headed the Department of Genetics at the Institute of Plant Industry (Leningrad, Soviet Union; before 1930, Russian Institute of Applied Botany and New Crops) and was Chair of Plant Genetics at Leningrad State University; in 1941, he was arrested for political reasons, together with several leading scientists, following the arrest of N. I. VAVILOV, director of VIR (Leningrad, Soviet Union), and later he was executed by firing squad; he was the first to produce polyploid hybrids between *Raphanus sativus* × *Brassica oleracea*.

KEMENESY, E. (1891-1981): he studied at the Agricultural Academy, Debrecen (Hungary), under Prof. K. KERPELY; as a soil scientist, he emphasized soil fertility for optimal plant production; he organized the Agricultural Research Institute at Keszthely and educated many noted Hungarian agronomists and plant breeders.

KERPELY, K. >>> KEMENESY, E.

KHVOSTOVA, V. (1903-1977): a famous Russian geneticist born in Moscow; after graduation from Moscow University, she worked as assistant to the Chair of Biology, the famous biologist V. F. NATALI; in 1935, she took postgraduate courses at the Institute of Cytology, Histology and Embryology where the famous geneticist N. K. KOLTSOV was director; her supervisor was N. P. DUBININ; she conducted cytogenetic studies on the position effect of the cubitus interruptus (*ci*) locus of *Drosophila melanogaster* and was simultaneously involved in studies on X-ray mutagenesis and mechanisms of chromosomal aberrations; after 1956, she focused her interest on plant mutagenesis; in 1965, she moved to Novosibirsk, where she organized the Laboratory of Cytogenetics in the newly formed Institute of Cytology and Genetics of the U.S.S.R. Academy of Sciences; under her guidance, cytogenetic and genetic contributions were made to wheat improvement; she became well known as the author of monographs such as *Cytogenetics of Wheat and Its Hybrids* (1971), *Potato Genetics* (1973), *Pea Genetics and Breeding* (1975), and *Cytogenetics of Mutations and Karyotype Evolution* (1977).

KIHARA, H. (1904-1986): he graduated from Hokkaido University (Japan) in 1918; the simultaneous discovery of the ploidy evolution of wheat by T. SAKAMURA and K. SAX drew his interest to the field; as professor of genetics at the Botany Department at the Kyoto University from 1920, he developed his wheat research; after two years of study at the Kaiser Wilhelm Institut für Biologie in Berlin (Germany) under C. CORRENS, he returned to Kyoto; he revealed step by step the secrets of wheat evolution from genome to plasmon interrelationships; he was one of the discoverers of *Aegilops squarrosa* as the donor of the third wheat (D) genome.

KISS, A. (1916-2001): born in Budapest (Hungary); he graduated from the Horticultural High School; at the age of eighteen, he moved to the Agriculture University Mosonmagyarovar (Hungary) to study Agricultural Sciences, where he earned a degree in agronomy; after working at a private farm (1936-1939), he began his research career at the Plant Breeding Institute, Mosonmagyarovar, in 1941; from 1950 to 1957 he worked at the Plant Breeding Institute of the Hungarian Academy of Sciences; his interest was focused on the genetics and breeding of pea, watermelon, and small-grain cereals, particularly wheat–rye hybrids; worked with G. REDEI, who emigrated to the United States in 1956 [they observed a strong genotypic control of crossability of various wheat and rye genotypes; their fundamental work was successful in developing primary octoploid triticale stocks]; he discovered early on that hexaploid forms are more useful in agriculture than octoploids; by 1962, he developed "No 30," the first secondary hexaploid triticale from a cross of "F481" (*T. aestivum* × rye // *T. turgidum* × rye); in 1968, his nursery released the world's first triticale cultivars for commercial production, as "T-No 57" and "T-No 64"; he promoted modern triticale breeding since they are as competitive on marginal soils as rye and contain 30-50 percent more protein; between 1967 and 1970, KISS developed more than ten hexaploid lines and sent them to the Cereal Research Institute, Szeged, for further tests on various locations; in 1970, his semidwarf cultivar "Bokolo" was patented in Germany; KISS received his PhD on the topic of "Microevolution of wheat–rye hybrids" in 1964; in 1972, he earned the DSc of the Hungarian Academy of Sciences for his work titled "Genetics and Breeding of Triticale"; he published more than 138 scientific papers.

KNIGHT, T. A. (1759-1838): from 1811 to 1838 he was the president of the Horticultural Society of London (England) as a botanist showing experience and scientific instinct; he was convinced that increased yields in plants and animals can be achieved by cross-breeding; in 1779, he emphasized the practical aspects of hybrids in grape, apple, pear, and plums, particularly to improve winter hardiness; his pea crosses (1799-1823) are of genetic interest; as did G. MENDEL, he recognized the advantage of pea as a research and breeding subject; he often noted the luxuriance of hybrids and the advantage of outcrossing to produce new forms; he was first to describe the dominance of the gray seed color over the white one, but did not calculate the relation of different segregating fractions as MENDEL did.

KOCH, K. H. E. (1809-1979): a German botanist who named and studied many of the plants known as cultivars; he brought great clarity and unity to the nomenclature of plants, cultivated and not cultivated.

KOHLI, S. P. >>> ANDERSON, R. G.

KOL, A. K. >>> LYSENKO, T. D.

KÖLREUTER, J. G. (1733-1806): between 1760 and 1766, this German carried out the first series of systematic experiments in plant hybridization using tobacco (*Nicotiana paniculata* × *N. rustica*); he demonstrated that the hybrid offspring generally resembled the parent as closely as the seed parent—thus, for the first time, he showed that the pollen grain had an important part in determining the characters of the offspring [this was a novel idea and was disbelieved by his contemporaries]; he also observed accurately the different ways in which the pollen can be naturally conveyed to the stigma of the flower and discovered the function of nectar and the role of wind in flower pollination; he observed that hybrid plants often exceed their parents in vigor of growth (hybrid vigor; now called "heterosis").

KOLTSOV, N. K. >>> KHVOSTOVA, V.

KÖNNECKE, G. (1908-1992): a German agronomist working as a professor at the University of Halle/Saale (Germany); he was concerned about utilizing crop rotations for increased agricultural productivity and choosing suitable crop varieties for different environments.

KORIC, M. (1894-1977): he was head of the Plant Breeding Station at Agricultural School at Krizevci (Serbia) from 1922 to 1929; by applying combination breeding and crossing domestic wheats with imported varieties, which carried genes for certain desirable traits (lodging resistance, earliness, rust resistance, bread-making quality), he achieved important progress in local wheat breeding; he introduced earliness and shorter straw by introgression of Italian wheats carrying genes of the Japanese wheat "Akakomugi"; quality improvement was achieved by crossing with Canadian and Indian ("Calcutta Red") spring wheats [among the best-known and most-spread new wheats were the cultivars "K6" and "K9"]; at Osijek (1929-1948) he continued his work and developed a number of additional varieties, such as "Koric's Awnless," "Osjecka Sisulja," or "U-1" [very soon they spread to over 50,000 ha and became leading cultivars in wheat production in the western parts of former Yugoslavia until the early 1960s].

KOSTOV, D. (1897-1949): a Bulgarian geneticist and cytogeneticist; he studied agriculture and obtained a PhD in agriculture at the University of Halle/Saale (Germany); under the guidance of E. M. EAST at Harvard University, he studied the ontogeny, genetics, and cytogenetics of *Triticum* and *Helianthus* hybrids, as well as tumors and other malformations on certain *Nicotiana* hybrids; later, at VAVILOV's laboratory of genetics, he continued research on polyploidy of crop plants.

KRISTEV, K. K. (1912-1986): an outstanding Bulgarian plant pathologist born in Khaskovo; he was educated at Sofia State Agro-Forestry Institute and trained as a young agronomist in 1935 at the newly established Institute of Plant Protection (Sapareva Banya); afterward, he served for about eleven years at the Department of Plant Pathology at the University of Sofia; he earned a PhD on smut and bunt diseases of wheat in 1943, and until 1976, he remained in different positions on this site; thereafter, he was appointed head of the International Wheat Immunity Laboratory at the Dobroudja Wheat and Sunflower Institute, General Toshevo/Varna (1976-1979); besides his teaching and organizing activities, his scientific activity was dedicated to research on smut and rust diseases, on the mechanism of enzyme activity, virology and toxicology, and on several new diseases of crops cultivated in Bulgaria.

KROLOW, K.-D. (1926-): professor of genetics and breeding research at the Technical University Berlin (Germany); most of his time he spent on the development of wheat–rye hybrids by sexual combination; besides the predominant octoploid and hexaploid triticales, he was the first to produce a viable tetraploid triticale by subsequent introgression of D genome chromosomes of hexaploid wheat.

KRONSTAD, W. E. (1932-2000): he was born in Bellingham (United States); he attended Washington State University, receiving a BSc degree in agronomy in 1957; in 1959, he was awarded an MSc degree in plant breeding and genetics and then joined the USDA wheat-breeding program at Washington State University as a research assistant with Dr. O. A. VOGEL; from 1959 to1963, KRONSTAD served as an instructor in the Farm Crops Department at Oregon State University and received his PhD degree in 1963; he remained at Oregon State University and was appointed project leader for cereal breeding and genetics in 1963.

KUCKUCK, H. (1903-1992): a German plant breeder born in Berlin; from 1925, he studied at the Landwirtschaftliche Hochschule, Berlin, and received his PhD in 1929; he started his professional career as an assistant at the Kaiser Wilhelm Institut für Züchtungsforschung at Müncheberg (Germany); for political reasons, he was released in 1936; in 1946, he was appointed to the Chair of Plant Breeding at the University Halle/Saale and returned to Müncheberg in 1948; he moved to Sweden as a guest researcher in 1950; from 1952 to 1954, he worked as a breeding expert and consultant in Iran on behalf of FAO; in 1954, he received a call from the Technische Hochschule, Hannover (Germany) and became full professor at the Institute of Horticultural Breeding, which he renamed the Institute of Applied Genetics; from this institute he retired in 1969; he was author of many scientific papers and several outstanding textbooks of plant breeding.

KÜHN, J. (1825-1910): a German agronomist and university teacher (Photo 6.1), he was the founder and the organizer of the first university study of agricultural sciences in Germany; he was born at

PHOTO 6.1. Julius Kühn (1825-1910) in his office. *Source:* R. Schlegel.

Pulsnitz (Germany) and studied at the Polytechnic School in Dresden before he became an agricultural volunteer in a manorial estate to learn the practice of agriculture; from 1848 to 1855, he was director of the estate Gross-Krausche/Bunzlau; there he studied crop plant diseases by advanced techniques (microscopy) and published several papers; in 1855, he became a student at the Agricultural School at Bonn-Poppelsdorf; he received his PhD from the University of Leipzig in 1857 (thesis: About the Smut Diseases in Cereals); in 1858, he published the book *The Diseases of Crop Plants, Origin and Protection,* which was famous as a textbook for many years; in 1862,

KÜHN became full professor of agriculture at the University Halle/
Saale; in 1863, he received governmental permission to establish an
independent Institute of Agriculture including lecture halls, experimental fields, and laboratories, which was developed over forty years
as the most important place of agricultural research and teaching in
Germany; in 1878, KÜHN began a permanent field trial with rye that
still continues after more than 125 years ("Ewiger Roggenbau," i.e.,
everlasting cultivation of rye, or a long-term testing experiment in
rye); in 1889, he founded the first Experimental Station for Nematode
Removal in order to prevent the so-called "beet exhausting" of soils;
he initiated >>> K. von RÜMKER's basic studies on cereal breeding; his most famous students were >>> E. von TSCHGERMAK-
SEYSENEGG, F. von LOCHOW, and >>> W. RIMPAU.

KULPA, W. (1923-1984): born in Bialobrzegi (Poland); from 1945
to 1950, he studied at the University of Lublin, and he started his career as seed scientist at the same university; in 1954, he became head
of the Chair of Plant Breeding and Seed Science of the University of
Lublin; he obtained his PhD in 1959 ("Biology and Germination of
Adonis vernalis L."); after a postdoctoral study at the Zentralinstitut
für Kulturpflanzenforschung, Gatersleben (Germany), he devoted
himself to the studies of systematics of *Linum usitatissimum* and later
of *Veronica* species; after twenty-two years of teaching and scientific
work at the University of Lublin, he was appointed as the head of the
Department of Plant Collections at the Institut Hodowli i Kalimaty-
zacji Roslin in Radzikow (Poland)—in that position he became the
main organizer of crop plant collections in Poland; he successfully
developed the scientific exchange with the European Association for
Research on Plant Breeding (EUCARPIA), the Food and Agriculture
Organization (FAO) of the United Nations, and the International Board
of Plant Genetic Resources (IBPGR); he organized and participated
in several expeditions within Poland [the Tatra mountain region,
Bieszczaden, Caucasian region] and in Turkey in order to collect wild
species of crop plants and safeguard plant genetic resources.

KUSH, G. S. (1942-): he is one of the global leaders on crop breeding and one of the major brains behind the development of productive
rice varieties and the "Green Revolution" in crop improvement; he

was born in Rurkee, Punjab (India); as the son of a farmer, he received his BSc from the Punjab Agriculture University and went to University of California, Davis (United States) to do his PhD at the age of twenty-five; in 1967, after postdoctoral studies on tomato breeding, KUSH was at the International Rice Research Institute (IRRI), Manila (Philippines) until 2000; for thirty-five years, he and his team introduced rice varieties like "IR8," "IR36," "IR64," and "IR72"; IRRI rice varieties and their progenies are planted in over 70 percent of the world's rice fields; he developed the so-called "miracle rice," "IR36," by using "IR8" as donor parent and crossing it with thirteen other parental varieties from six countries ["IR36" is a semidwarf variety that has proved highly resistant to a number of the major insect pests and diseases; it matures rapidly in about 105 days as compared with 130 days for "IR8" and 150-170 days for traditional types; the combination of suitable characteristics made "IR36" one of the most widely planted food crop varieties of the world ever]; in 1994, he announced a new type of "super rice," which has the potential to increase yields by 25 percent; KUSH's latest work deals with the so-called "new plant type" for irrigated rice fields.

LAUBSCHER, F. X. (1906-): he was born in Vredenburg (South Africa) and received his education there; he qualified for the BSc in 1928, MSc in 1942, and DSc in 1945; after fulfilling various research and teaching assignments in various parts of the country, most of these on behalf of high-level government agencies, he was appointed to the Chair of Genetics at his *alma mater,* Stellenbosch University (South Africa), in 1950; from 1936 to 1949, while stationed at the Potchefstroom College of Agriculture, he was involved in wheat breeding, and he released the cultivars "Spitskop," "Goudveld," "Magaliesburg," and "Flameks" (a wheat with exceptional baking attributes); he published a monograph entitled *A Genetic Study of Sorghum Relationships;* in 1949, he was appointed as technical advisor to the Maize Board, in which, together with scientists from the United States, he was responsible for the successful initiation of the national maize-hybrid-breeding program.

LEDERBERG, J. >>> BEADLE, G. W.

LEEUWENHOEK, A. van (1632-1723): a Dutch microscopist, called "Father of Scientific Microscopy," he was a cloth merchant and wine taster by trade; his works were published under the title *Secrets of Nature* (1668); the bacteria he found for the first time underthe microscope he described as "animacules" (little animals); LEEUWENHOEK extended MALPIGHI's demonstration of blood capillaries, and six years later gave the first accurate description of red blood corpuscles, building on W. HARVEY's (1578-1657) discovery of the circulation of the blood in 1628; he discovered the effect of aphids on plant life and showed that they reproduce parthenogenically; he described different stem structures in monocots and dicots, and observed *polyembryony* in citrus seed; he was responsible for the first representation of bacteria in a drawing in 1683; he constructed over 400 microscopes and bequeathed 26 to the Royal Society of London.

LEIN, A. (1912-1977): a distinguished German cereal breeder; he started his career at the Cytogenetic Department of the Institute of Plant Cultivation and Plant Breeding of the University of Halle/Saale (Germany); in 1942, he completed his PhD on the subject "The genetic basis of the ability of crossbreeding between wheat and rye"; starting in 1944, he evaluated wheats from collections of the German Hindu Kush expedition of 1935-1936; in 1947, he continued his scientific work as the head of the Department of Self-Fertilization at the Max Planck Institut für Züchtungsforschung at Voldagsen (Germany); in 1949, he became the head breeder of the company Ferdinand Heine at Schnega (Germany); in 1969, he became the head breeder for barley and wheat at F. von Lochow-Petkus Ltd. (Germany) [among his numerous successful varieties are the winter wheats "Kranich" and "Kormoran," the spring wheats "Kolibri" and "Selpek" and the spring barley "Oriol"].

LELLEY, J. (1909-2003): born in Nyitra (Hungary); he was one of the outstanding Hungarian wheat breeders (Photo 6.2); he studied in Budapest; in 1931, he graduated from the Agricultural High School at Mosonmagyarovar; after a short stay at Bratislava (Slovakia) in an agricultural office, he received postgraduate training in plant breeding after 1946; he moved to the Plant Breeding Station at Kompolt and

PHOTO 6.2. Janos Lelley (1909-2003) in late age. *Source:* T. Lelley, Tulln (Austria).

developed there an extensive wheat-breeding program and conducted basic research on the methodology of breeding and on resistance to different stresses; he organized a special network of research stations in the mountains for screening wheat for frost resistance; he achieved improvements of leaf rust and drought tolerance of wheat and worked out an effective method for artificial rust infection; he bred the spring and winter wheat "K169"; in 1962, he started to organize a new wheat breeding center at Kiszombor, and there he improved the cultivars "Kiszombori 1" and "GK Tiszataj" [the latter was one of the best quality wheats in Europe with a protein content of 16-17 percent]; in 1954, he wrote a handbook on wheat breeding, *Wheat Breeding—Theory and Practice,* for Hungarian students and breeders.

LEVAN, A. >>> MÜNTZING, A.

LIEBIG, J. von >>> DELBRÜCK, M.

LINDLEY, J. (1799-1865): British scientist, one of the most re-markable horticultural scientists of the nineteeth century; his book *The Theory of Horticulture* (1840) is a classic and is still considered "one of the best books ever written on the physiological principles of horticulture"; his formal education lasted only to the age of sixteen, but his affinity for hard work enabled him to become one of the most productive plant scientists of his era; LINDLEY had several careers, most of them simultaneously: he was the "mainspring" of the London Horticultural Society for forty years, professor of botany at the University of London for thirty-three years, editor of the *Botanical Register* for eighteen years, editor of the *Gardener's Chronicle* for twenty-five years, and professor of botany and director of the Physic Garden; he played a major role in saving Kew Gardens from being disbanded by the government as a budget-cutting measure; his pioneering works on orchid taxonomy earned him the title of "Father of Modern Orchidology"; he authored books on medical uses of plants, general botany, popular horticulture, and fossil plants; his botanical texts helped establish the natural system of plant classification as the system of choice; he named innumerable new species brought back by plant explorers, and started the practice of ending plant family names in "aceae"; as editor for the *Gardener's Chronicle* for twenty-five years, he worked to improve the state of horticultural science; his book *Theory of Horticulture* (1831) had a major impact on horticulture, and was translated into German, Dutch, and Russian and published in an American edition (1841).

LINNAEUS, C. (Carl von LINNÉ) (1707-1778): a Swedish botanist and physician; he began his education as a theology student, but at the age of twenty-three he became curator of the Gardens of the University of Lund (Sweden); from 1732 to 1738, he traveled through Lapland, Holland, England, and France, then returned to Stockholm, where he practiced medicine; in 1741, he became the head of the Botany Department at the University of Uppsala, where he remained until his death; his botanical contributions have earned him the title of "Father of Taxonomy;" LINNAEUS established groups of organisms, large and small, based on structural or morphological similarities and differences [the basic taxonomic criteria for grouping plants was based on the morphology of their reproductive parts—the plant organs least likely to be influenced by environmental conditions]; his "sexual system" of classification used the number of stamens and carpels (styles) as a method of grouping plants (an artificial system, however, that is no longer used); he described and assigned names to more than 1,300 different plants; he is credited with the establishment of the binomial nomenclature, and with having replaced the long-winded and confusing descriptions of the herbalist with clean and succinct description; his works include *Systema Naturae* (1735), *Fundamenta Botanica* (1736), *Genera Plantarum* (1737), *Classes Plantarum* (1738), and *Philosophia Botanica* (1751).

LOCHOW, F. von (the third of LOCHOW dynasty, 1849-1924): a German agronomist who conducted his breeding on his estate at Petkus; when he selected the first seeds from the Austrian landrace "Probsteier Winterroggen" in 1880, he recognized the wide variability between the single-seed progenies, which inspired him to start breeding with rye [the subsequent elite breeding improved the winter hardiness, fertility, and uniformity of seeds]; in 1925, >>> RÜMKER described his method as subsequent pedigree breeding (independently established from VILMORIN in France) and FRUWIRTH (1908) as a "German selection method"; the first yield testing of the German Agricultural Society in 1891 showed that "Petkuser Winterroggen" (Petkus winter rye) outyielded standard ryes by 8 to 11 percent; from that time "Petkus rye" (1891-1960) was distributed around the world and served as basic gene pool for many other rye varieties; in 1911, a spring type of this rye was released too, and multiplied until 1960 as "Petkuser Sommerroggen."

LUKYANENKO, P. P. (1901-1973): a very successful Russian wheat breeder working at Krasnodar; he became known throughout the world in connection with outstanding varieties such as "Bezostaya 1"; [it became popular in every wheat-producing country of the world]; his varieties were grown on more than 10 million ha; new varieties followed, such as "Avrora," "Kavkaz," "Bezostaya 2," "Skorospelka," or "Rannaya 12"; the varieties "Avrora" and "Kavkaz" carried the so-called 1RS.1BL translocation and were transferred around the world—they are still used in many wheat improvement programs.

LURIA, S. >>> DELBRÜCK, M.

LYSENKO, T. D. (1898-1976): a Russian botanist and horticulturist; he was born at Karlowka (Ukraine); after studying at the Horticultural School Belozersk (1921) and the Agricultural Institute of Kiev (1925), he was appointed to the breeding station at Gandže (Azerbaidshan) from 1925 to 1929; LYSENKO was a young agronomist when he came learned of an experiment in the winter planting of peas to precede the cotton crop in the Transcaucasia region, and subsequently became famous for the discovery of "vernalization," an agricultural technique that allows winter crops to be obtained from summer planting by soaking and chilling the germinated seed for a determinate period of time; he was the first to use the term "vernalization" but was not, in fact, the first to discover this technique; LYSENKO ignored previous studies of thermal factors in plant development; after being overshadowed by MAKSIMOV, his critic, at the All-Union Congress of Genetics, Selection, Plant and Animal Breeding held in Leningrad in 1929, LYSENKO organized a boisterous campaign around vernalization and made extravagant claims based on a modest experiment carried out by his peasant father; the Ukrainian Commission of Agriculture, in the hope of raising productivity after two years of famine, ordered massive use of the vernalization technique; LYSENKO was moved to a newly created department for vernalization at the All-Union Institute of Genetics and Plant Breeding in Odessa, and there he began to publish the journal *Jarovizatsiya (Vernalization),* in which he disseminated his ideas on a wide scale and created a mass movement around vernalization; the next stage in LYSENKO's career came when, from 1931 to 1934, he began to advance a theory to explain his technique; according to the idea of the phasic development of plants—a plant underwent various stages of development, during each of which its environmental requirements differed sharply; the conclusion LYSENKO drew from this was that knowledge of the different phases of development opened the way for human direction of this development through control of the environment; it was a very vague theory, never to be spelt out very fully, but it provided the link in the evolution of LYSENKO's platform from a simple agricultural technique to a full-scale biological theory; the underlying theme was the plasticity of the life cycle; LYSENKO came to believe that the crucial factor in determining the length of the vegetation period in a plant was not its genetic constitution but its interaction with its environment; his theory developed in a pragmatic and intuitive way a rationalization of agronomic practice and a reflection of the ideological environment surrounding it and not being pursued according to rigorous scientific methods; contemporaneous with LYSENKO's vernalization movement was a growing interest in the work of >>> I. V. MICHURIN (1855-1935), the last in the line of an impoverished aristocratic family in central Russia, who cultivated fruit trees and began experimenting with grafting and hybridization; MICHURIN worked on the assumption that the environment exercised a crucial influence on the heredity of organisms, and he queried the relevance of MENDEL's so-called "peas laws" to fruit trees; MICHURIN's name was soon to be seized upon by LYSENKO to designate a whole new theory of biology in opposition to classical genetics, even though MICHURIN himself had no such theoretical pretensions; in 1931 and 1932, a number of geneticists were branded as "menshevising idealists" and lost their positions at the Communist Academy of Sciences; a particularly vicious article that appeared in the newspaper *Ekonomicheskaya Zhizn* in 1931 was directed against Academician N. I. VAVILOV, founder and president of the Lenin Academy of Agricultural Sciences, director of its All-Union Institute of Plant Breeding, as well as director of the Institute of Genetics of the Academy of Sciences; VAVILOV was an internationally eminent plant geneticist and an ardent advocate of the unity of science and socialism; the article was written by a subordinate of VAVILOV, A. K. KOL, who accused VAVILOV of a reactionary separation of theory and practice and advised him to stop collecting exotica and to concentrate on plants that could be introduced directly into farm production; unrealizable goals were imposed on VAVILOV's All-Union Institute of Plant Breeding in 1931, and in 1934 he was called in by the Council of Peoples Commissars to account for the "separation between theory and practice" in the Lenin Academy of the Agricultural Sciences; LYSENKO was very much a part of this campaign, stirring up a negative attitude to basic research and virulently demanding immediate practical results; in 1940, VAVILOV was arrested, and LYSENKO replaced him as director of the Institute of Genetics of the Academy of Sciences; he believed in Lamarckianism, and by his positions (as president of Agricultural Academy, 1938-1956, and director of the Institute of Genetics of the Academy of Sciences, 1961-1962) he negatively influenced agricultural policy in the former Soviet Union.

MACKEY, J. >>> NILSSON-EHLE, H.

MALINOWSKI, E. >>> BOJANOWSKI, J. S.

MALPIGHI, M. (1628-1694): he was an Italian physician, anatomist, physiologist, and a pioneer microscopist; he graduated in medicine in 1653, became a lecturer in 1656, and was appointed to the Chair of Theoretical Medicine at the University of Pisa (Italy); he became the personal physician to Pope INNOCENT XII; his major contribution was *Anatome Plantarum* (1675); he was one of the first to utilize the microscope in the study of animal and vegetable structures and is considered the founder of microscopic anatomy; he applied himself to vegetable histology and became acquainted with spiral vessels of plants in 1662; he made the important discovery that the layers of tissues in leaves and young shoots are continuous with those of the main stem; he distinguished fibers, tubes, and other constituents of wood, and was the first to understand the food functions of leaves; he observed stomata in leaves and nodules on legume roots and realized that the ovule developed into a seed and the carpel into a fruit or a portion thereof.

MARAIS, G. F. >>> PIENAAR, R. de V.

MÄRKER, M. (1842-1901): a German agronomist who promoted plant breeding at the agrochemical research station Halle/Saale (Germany); he introduced a system for testing crop varieties for specific environments.

MARSCHNER, H. (1929-1996): he was born at Zuckmantel (Czech Republic); he studied agriculture and chemistry at the University of Jena (Germany), obtained a PhD in Agricultural Chemistry in 1957, and then joined the Institut für Kulturpflanzenforschung, Gatersleben (Germany); during these years, he developed his interest in modern techniques for studying plant nutrient uptake; in 1966, he became professor of plant nutrition at the Technische University of Berlin, and from 1977 he was the director of the Institut für Pflanzenernährung at the University of Hohenheim; he was one of the most highly esteemed scientists in the area of plant mineral nutrition; at the beginning of his career he mainly studied the uptake of mineral nutrients, but then he extended this to include nutrient transport and use within the plant; his later research greatly advanced the understanding of rhizosphere processes and iron uptake by plants; he also included environmental aspects of plant nutrition in his work, e.g., on the side effects of high rates of agricultural fertilizer use, on heavy-metal contamination of soils, and on the effect of changes in forest ecosystems on the uptake and use of nutrients by trees; he published extensively on the adaptation mechanisms of plants to adverse soil conditions and low nutrient supply; studies on the efficient use of fertilizers in developing countries were of particular importance to him, in recent years especially in Turkey, West Africa, and China; he was one of the first to relate plant nutrition phenomena with genetic control and breeding approaches.

MARTYN, T. >>> MILLER, P.

MAYSTRENKO, O. I. (1923-1999): a prominent Russian geneticist; expert in cytogenetics; she was born in Orsk (Urals) into a family of Ukrainian farmers; she finished school in Samarkand (Uzbekistan) and studied at the Moscow Timiryazev Agricultural Academy (1942-1947); in 1947, after graduation from the Department of Breeding and Seed Multiplication, she began her scientific career as spring and winter barley breeder at a breeding station in Kirgizia, her work in this period resulting in the cultivar "Nutans 45"; from 1950 to 1954 she took a postgraduate course in the All-Union Institute of Plant Industry in Leningrad (VIR), where she gained her PhD; she began to work with wheat firstly in Kirgizia and later in Sverdlovsk, where she became the head of the Cereal Laboratory (1951-1960); she developed breeding on fast ripening cultivars of oats and wheat with high bread-making quality; her major scientific contributions followed after her appointment in 1960 to the Institute of Cytology and Genetics of the Siberian Branch of the Russia Academy of Sciences at Novosibirsk; having been acquainted with the cytogenetic stocks of >>> E. R. SEARS, she utilized them for the development of adapted wheats for Russia; she developed the monosomic series and later ditelosomics as well as monotelosomics of Russian cultivar varieties "Diamant" and "Saratovskaya 29" [the choice of these two varieties is evidence of her foresight; "Diamant" is an outstanding genotype of high-grain protein, and "Saratovskaya 29" is superior in bread-making quality and adaptive potential]; she contributed to the chromosomal localization of genes *Vrn1* and *Vrn3* and a gene for resistance to race 20 of leaf rust as well as genetic factors determining physical properties of dough, resistance to lodging and plant height; she established various sets of intervarietal substitution and near-isogenic lines (NILs); she also successfully transferred disease resistance to the wheat cultivar "Saratovskaya 29" by crossing with the synthetic hexaploid wheat *T. timopheevii* the *T. tauschii.*

McCLINTOCK, B. (1902-1992): born in 1902 at Hartford, Connecticut; she studied plant genetics at Cornell University in Ithaca, New York, receiving her doctorate in botany in 1927; she took a research position at Carnegie's Cold Spring Harbor Laboratory (New York), which she held for more than forty years; by observation and experimentation with variations in the coloration of kernels of maize, she discovered that the genetic information is not stationary and suggested that the transposable elements were responsible for the diversity in cells during an organism's development; she won the Nobel Prize for Physiology in 1983 for the discovery of mobile genetic elements, a discovery that heavily influenced molecular genetics during the last two decades of the twentieth century.

McEWAN, M. (1931-2004): New Zealand's foremost wheat breeder, from Palmerston North; his achievements helped change bread from the white, unsliced loaves of the 1960s and early 1970s to the conveniently sliced, multigrain loaves of today; while working for the Department of Scientific and Industrial Research (DSIR), Crop Research (Wellington), he bred several highly successful wheat cultivars, including "Otane" [which had excellent milling qualities and produced exceptionally high-grade flours; for a period in the early 1990s "Otane" commanded over 80 percent of New Zealand's wheat production]; other successful wheat cultivars, that he bred he named after areas in the Manawatu; "Rongotea" (1979), "Oroua" (1979), "Karamu" (1972), and "Endeavour" (1994) [all these cultivars resulted from semidwarf wheat germplasm he brought to New Zealand]; McEWAN worked on other cereals too, releasing the successful general purpose feed oat "Awapuni," a forage oat "Enterprise" for Australia, the black feed oat "Finlay," a forage barley "Opiki," and the triticale "Aranui."

MEITNER, L. >>> DELBRÜCK, M.

MELCHERS, F. (1905-1997): a German geneticist; he strongly contributed to cell research and utilization of cell techniques in breeding; beginning with mutation research, he was the first who developed (in the early 1980s) a vital potato–tomato hybrid by somatic cell fusion of protoplasts, the so-called "Tomoffeln" or "Karmaten."

MELLO-SAMPAYO, T. (1923-1997): he was born at Pangim, Nova Goa, ex-Portuguese state of India; he graduated in 1949 in Agronomy from the Technical University of Lisbon (Portugal); after a short period in Mozambique, he went to the National Agricultural Station, where he began his cytogenetic studies under the supervision of Prof. Antonio CAMARA, mainly on wheat (e.g., aneuploidy, chromosome pairing regulation and nucleolus-organizer activity); he paid particular attention to aneuploids of tetraploid wheat and obtained two substitution lines ("Camara" and "Resende"); he studied the dose effects of the *Ph* gene on chromosome pairing, achromatic fusion, and chromosome interlocking; with other collaborators he developed the practical and theoretical concept of mixoploid genomes of wheat and triticale; he devoted several papers to the regulation of NORs and amphiplasty in interspecific hybrids.

MENDEL, G. J. (1822-1884): an Austrian Augustinian monk in the monastery of Brünn (now Brno, Czech Republic); through experimentation, he discovered the underlying principles of heredity based on his work with pea plants, but his work was so brilliant and unprecedented at the time that it took thirty-four years for the rest of the scientific community to catch up to it; the short monograph *Experiments with Plant Hybrids,* in which he described how traits are inherited, has become one of the most enduring and influential publications in the history of science; MENDEL, the first person to trace the characteristics of successive generations of a living thing, was not a world-renowned scientist of his day; he was the second child of a farmer in Brünn; MENDEL's performance at school as a youngster encouraged his family to support his pursuit of a higher education, but their resources were limited, so MENDEL entered an Augustinian monastery, continuing his education and starting his teaching career; his attraction to research was based on his love of nature; he was not only interested in plants but also in meteorology and theories of evolution; MENDEL often wondered how plants obtained atypical characteristics; on one of his frequent walks around the monastery, he found an atypical variety of an ornamental plant, and he took it and planted it next to the typical variety; he grew their progeny side by side to see if there would be any approximation of the traits passed on to the next generation; this experiment was "designed to support or to illustrate LAMARCK's views concerning the influence of environment upon plants;" he found that the plant's respective offspring retained the essential traits of the parents, and therefore were not influenced by the environment; this simple test gave birth to the idea of heredity; he saw that the traits were inherited in certain numerical ratios; then he came up with the idea of dominance and segregation of genes and set out to test it in peas; it took seven years to cross and score the plants in their thousands to prove the laws of inheritance; from his studies, MENDEL derived certain basic laws of heredity, such as that hereditary factors do not combine but are passed intact, that each member of the parental generation transmits only half of its hereditary factors to each offspring (with certain factors "dominant" over others), and that different offspring of the same parents receive different sets of hereditary factors.

MERKLE, O. G. (1929-1999): he was born in Meade (United States); he earned a BSc in agronomy in 1951 and an MSc in plant breeding in 1954, both degrees from Oklahoma State University; in 1957, he moved to College Station, Texas, where he worked as an agronomist with the USDA; he received his PhD from Texas A&M University in 1963, with a major in plant breeding and a minor in plant pathology; he continued with the Agricultural Research Station (ARS) at College Station until 1974, when he transferred to Stillwater, Oklahoma; there, he worked as an ARS research agronomist until he retired in 1988; his work encompassed many facets of practical research, ranging from interactions of the environment with fertilization and plant spacing to the inheritance of wheat flour quality and resistance to pests and drought; improvement and development of small-grain germplasm and cultivars were among his most important contributions to agriculture; as agronomist-breeder on research teams in Texas and Oklahoma, he made significant contributions to the development of flax cultivars "Caldwell," "Dillman," and "Mac," and wheat cultivars "Caddo," "Milam," "Sturdy," "Fox," "Mit," and "Century"; he and his co-workers released and registered more than fifteen wheat germplasm lines with resistance to disease (rust) or insects (Hessian fly and the yellow sugarcane aphid) and with improved characters (large seed), and pearl millet germplasm lines with resistance to chinch bug; prior to retirement, he was active in evaluating barley, wheat, and wild *Triticum* species for resistance to the Russian wheat aphid; he also evaluated winter and spring wheat genotypes for tolerance to drought in cooperative programs with colleagues at Lubbock (United States) and El Batan (Mexico).

METTIN, D. (1932-2004): he attended Berlin Eosander Junior College from 1942 to 1943, Eisleben Martin Luther Gymnasium from 1943 to 1950, Salzmünde Agricultural College from 1950 to 1952, and University of Halle/Saale from 1952 to 1955, where he obtained an MSc; he obtained his PhD from the same University in 1961, with a major in plant breeding and genetics under the supervisor >>> Prof. H. STUBBE (thesis: Genetic and Cytological Studies in the Genus *Vicia*); he was awarded a DSc degree from that institution in 1977 (thesis: Selection, Identification and Genetic Utilization of Aneuploids in Hexaploid Winter Wheat, *Triticum aestivum* L.); in 1961, he moved to the Institute of Plant Breeding at the University of Halle/Saale, where he began his professional career after graduation as assistant professor at the Institute of Plant Breeding, Hohenthurm; in 1968, he was appointed as reader of cytogenetics in the Institute of Plant Breeding, and in 1977, as full professor of plant breeding; in 1983, METTIN accepted a call as director of the Central Institute of

Genetics and Crop Plant Research Gatersleben, from which he retired in 1991; during his time at the University of Halle/Saale he contributed significantly to the improvement of academic education in
genetics, plant breeding, and seed production as well as applied cytogenetic research; he had the opportunity to encourage and direct several graduate students [more than forty BSc and more than sixty MSc
students studied under his supervision; many of the research projects
involved the participation of his twelve PhD and two DSc appli-
cants]; METTIN's research work encompassed many facets of basic
and applied research, including cytotaxonomic studies of *Vicia*, induced auto- and allopolyploidization in *Brassica* and *Secale*, wide
hybridization in *Triticineae*, production and utilization of aneuploids
in *Aegilops*, *Secale*, and *Triticum*, genetic mapping of resistance genes
to leaf diseases, quantitative traits in wheat, homoeologous chromosome pairing in cereals, and first applications of molecular genetics
and biotechnology in plant breeding; he was among the pioneers of
wheat aneuploid research; perhaps his four greatest accomplishment
are the creation of the complete series of monosomics of the German
wheat cultivar "Poros" (winter type) and "Carola" (spring type), the
production of the first complete series of primary rye trisomics, the
codiscoveries of the 1B/1R wheat–rye translocations and substitutions in hexaploid wheat, and the spontaneous homologous recombination between wheat and rye chromosomes.

MIKUZ, F. (1889-1978): he was a co-worker of >>> F. JESENKO,
and from 1921, the head of the newly established Plant Breeding
Station in Beltinci (Yugoslavia); he was engaged in the breeding of
wheat, oats, rye, maize, and buckwheat; his wheat cultivars "Beltinska
227," "Beltinska 321," and "Beltinska 831" were known and widespread for a long time in Slovenia.

MILLER, P. (1691-1771): a British gardener to the Worshipful
Company of Apothecaries at their Botanic Garden in Chelsea (United
Kingdom), and known as the most important garden-writer of the
18th century; *The Gardener's and Florist's Dictionary and a Com-
plete System of Horticulture* (1724) was followed by a greatly im-
proved edition titled *The Gardener's Dictionary,* containing the
"Methods of Cultivating and Improving the Kitchen, Fruit and
Flower Garden" (1831); this book was translated into Dutch, French,

and German, and became a standard reference for a century in both England and America; in the seventh edition (1759), he adopted the Linnaean system of classification; the edition enlarged by Thomas MARTYN (1735-1825), professor of botany at Cambridge University (United Kingdom), may be the largest gardening manual to have ever existed; MILLER is credited with introducing about 200 American plants; the sixteenth edition of one of his books, *The Gardeners Kalendar* (1775), gives directions for gardeners month by month and contains an introduction to the science of botany.

MILOHNIC, J. (1920-1974): a Croatian breeder at the Institute of Breeding and Crop Production, Zagreb; his main concern was cereal and fodder legume breeding; the winter vetch variety "Ratarka" has been of great local importance.

MISCHER, F. (1844-1895): a German chemist; he first described a method for purification of nuclei from the cytoplasm; from the nuclei he isolated an acid compound, the "nuclein;" later it was called "nucleic acid."

MITSCHURIN, I. W. (1855-1935): a Russian botanist and plant breeder; born in Dolgoje near Rjasan (Russia, now Mitschrowka); he bred more than 300 varieties of fruit trees, grape vine, and berries; his varieties were particularly suited for the cold climate of Russia; not recognized under the Czar's regime, he received much attention from the Soviet government after 1917; his success was based on the so-called "mentor method," believing that modifications are induced by environmental factors not by genes; T. D. LYSENKO became later a propagandist of MITSCHURIN's ideas, misleading Soviet biological sciences for decades.

MORGAN, T. H. (1866-1945): an American geneticist working on the fruit fly *(Drosophila melanogaster)*, demonstrating that genes are located on chromosomes; along with >>> W. BATESON, the co-founder of modern genetics, an experimentalist and originally a mutationist opposed to natural selection, Lamarckism, orthogenesis, and the chromosomal theory of inheritance; converted to the chromosomal theory of inheritance and to Darwinism after 1910.

MULLER, H. J. (1890-1967): an American geneticist best remembered for his demonstration that mutations and hereditary changes can be caused by X-rays striking genes and chromosomes of living cells; attended Columbia University from 1907 to 1909; a laboratory assistantship in zoology in 1912 allowed him to do research on *Drosophila* at Columbia; produced a series of papers on the mechanism of crossing-over of genes; his PhD dissertation established the principle of the linear linkage of genes in heredity; awarded the Noble Prize for Physiology or Medicine in 1946.

MÜNTZING, A. (1903-1984): born in Göteborg (Sweden); he became interested in the work of H. NILSSON-EHLE when he was a student; the group around NILSSON-EHLE influenced his early concept of species when he was a teacher at the University of Lund (Sweden); he also studied the theory of Ö. WINGE on the role of polyploidy in plant evolution; in his PhD thesis in 1930 he reported about the polyploidy in the genus *Galeopsis;* he was able to resynthesize a Linnean species *Galeopsis tetrahit* ($2n = 4x = 32$) by combining *G. speciosa* ($2n = 2x = 16$) and *G. pubescens* ($2n = 2x = 16$); at this time MÜNTZING was serving as a sugar beet breeder at Hilleshög; however, in 1931 he founded a cytogenetic department at Svalöf in order to study and to utilize polyploidy as a factor in breeding; NILSSON-EHLE stated the reason for its establishment as follows: "In recent years geneticists have found that important changes in the structure of organisms may be induced by intentionally changing the actual number of chromosomes (and thereby the number of genes) in the nuclei. It seems to be especially important to try to induce an increase in the chromosome number, since such an increase is often connected with an increased size of the plant organs and consequently with the production of increased vegetative yield"(In *Svalöf 1886-1986, Research and Results in Plant Breeding*. Lts förlag, Stockholm, Sweden, 1986, p. 18); after some tentative experiments to produce polyploids, he found the colchicine method, discovered by A. F. BLAKESLEE in 1937, to be a general and very useful method; MÜNTZING was successful in producing autopolyploids such as tetraploid rye as well as allopolyploids such as triticale; he also continued his research on *Galeopsis, Lamium, Potentilla*, etc.; in 1993, he published a paper on apomictic and sexual seed formation in *Poa* and was the first to demonstrate apomictic seed formation in *Poa pratensis* [later >>> E. AKERBERG found both apomictic and sexual types in this species and that apomictic plants were of the apogamic type combined with pseudogamy]; when in 1938 >>> NILSSON-EHLE retired as professor of genetics at the University of

Lund, MÜNTZING succeeded him; due to his concept of genetics, his chair at the University of Lund (Sweden) covered a broader field of genetics than just crops—he incorporated human genetics and developed a chromosome research branch, which was headed by A. LEVAN; he was very happy to see the success of triticale breeding during 1970-1980s, which he had started during the 1930s.

NABA, K. (1595-1648): a Japanese *Prunus* expert and the first person to ever write a book devoted to it; he documented and described many cultivars for the first time.

NÄGELI, C. >>> CORRENS, C. E.

NATALI, V. F. >>> KHVOSTOVA, V.

NAVASHIN, S. G. (1857-1930): he was born in Tsarevshchina (Russia); in 1874, after finishing gymnasium in Saratov, he entered the Medico-Surgical Academy in St. Petersburg; in 1878, he realized that medicine was not the sphere of knowledge and activity to which he wanted to devote himself, and so he entered the Natural Sciences Faculty of Moscow University, where he continued to specialize in chemistry; his profound knowledge in this field was noticed by K. A. TIMIRIAZEV, who lectured on plant physiology at the university; the final master examination, which NAVASHIN passed successfully at St. Petersburg University, was on botany; it allowed him to lecture on mycology and phytopathology; he gave the introduction to fungi systematics; at the same time he began his first mycological investigations; when he investigated heterospory in sphagnous moss, NAVASHIN proved that the filiform structures in the so-called "female flower" of the *Sphagnum* species are hyphae of the new fungus species *Helotium shimperi* (Discomycetes); he demonstrated that microspores of these mosses are in fact spores of smut fungus *Tilletia sphagni;* he also described new fungi species, including *Gymnosporangium tremelloides* and *Puccinia wolgensis;* in 1889, he moved again to St. Petersburg University as an assistant to Prof. I. P. BORODIN, where his attention was drawn to the genus *Sclerotinia* on plants of the genus *Vaccinium* and the slime mold *Plasmodiophora brassicae*—agent of club root in cabbage; in 1894, NAVASHIN held the Chair of Morphology and Systematics of Plants of St. Vladimir University in Kiev; he began to thoroughly study *Plasmodiophora brassicae* and described the vegetative stage of this intracellular parasite and its sporogenesis; NAVASHIN for the first time focused his attention on the cytology of sporogenesis, and in particular he determined the phenomenon of simultaneous nuclei division in the plasmodium of *P. brassicae;* studying the relationships between the parasite and host plant, he demonstrated for the first time that the protoplasm of cabbage cells affected by amoebas of slime mould remains viable; after the discovery of double fertilization in angiosperms (1898), he dedicated himself to embryological and karyological investigation of higher plants, but mycology always remained in the sphere of his scientific interests; in 1913, because of the state of his health, he moved from Kiev to Tiflis (Georgia); in 1923, he was invited to Moscow for the post of director of the Timiriazev Institute.

NEETHLING, J. H. (1887-1960): he was born in Lydenburg, Mpumalanga (South Africa); during the Anglo Boer war, he accompanied his father on commando missions; after the war, he returned to the Boys High School at Stellenbosch to complete his studies, and matriculated in 1906; in 1907, he gained his BA from the Victoria College; he was one of a group of young scholars selected by General L. BOTHA to further their studies in agriculture overseas; in 1911, he was awarded a BSc by Cornell University; in the same year he completed his MSc under Profs. WEBBER and GILBERT; prior to returning to South Africa, he received tutelage in hybrid maize breeding under the guidance of Prof. MOORE in Wisconsin, and then was instructed in genetics by >>> Prof. R. GOLDSCHMIDT in München (Germany), as well as by H. de VRIES in Amsterdam (the Netherlands); the existence and well-being of the small-grain industry in South Africa was largely built on NEETHLING's knowledge and expertise; for almost a half century, the wheat cultivars grown in the South Western Cape Province emanated from his breeding program; the wheat cultivars "Union 17" and "Union 52" were released in 1915, followed by "Gluretty," "Hoopvol," "Koalisie," and especially "Pelgrim," "Vorentoe," and "Sterling"; in 1949, the Wheat Board built a greenhouse and adjoining laboratory at the Welgevallen Experiment Station, and named it, in 1982, in his honor at the Senate of the University of Stellenbosch Agriculture Building; during his tenure, four DSc and twenty-nine MSc degrees were conferred; NEETHLING retired in 1949, and was succeeded by one of his students, Prof. >>> F. X. LAUBSCHER.

NEWMAN, L. H. >>> GOULDEN, C. H.

NEWTON, I. >>> HOOKE, R.

NIEDERHAUSER, J. S. (1917-2005): an American potato scientist and 1990 World Food Prize winner of the University of Arizona; he was a pioneer in international cooperation for the improvement of agricultural productivity and was known for developing potato varieties; in 1946, NIEDERHAUSER joined the newly formed Rockefeller Foundation Mexican Agricultural Program; he spent fifteen years working in Mexico on maize, wheat, and bean production; during this time, he began to study potato production in Mexico; his work over the next decades focused on the improvement of potato production in many developing countries; due to the success of this work, the International Potato Center, now supported by the Consultative Group on International Agricultural Research (CGIAR), was established in Lima (Peru) in 1971; in 1978, he established the Regional Cooperative Potato Program in Mexico, Central America, and the Caribbean; one of NIEDERHAUSER's important scientific contributions was the development of potato varieties with resistance to late blight disease; he discovered that the source of the pathogen responsible for the Irish potato famine came from Mexico, and that many wild inedible potato species possessed a durable field resistance to the late blight fungus; NIEDERHAUSER's work resulted in the establishment of the potato as the fourth major food crop worldwide; as another result, potato production in Mexico increased from 134,000 t in 1948 to greater than 1 million t by 1982.

NILSSON-EHLE, H. (1873-1949): a Swedish botanist trained at the University of Lund; he was employed at the Swedish Seed Association, Svalöf, in 1888, where he was the director from 1890 to 1924; when the Svalöf Institute started in 1886 (cf. Table 3.1), the first director was a German from Kiel, T. BRUUN von NEERGAARD; his successor was an agricultural engineer by training, who constructed a number of instruments for measuring, analyzing, and sorting spikes and seeds of cereals; the German >>> RÜMKER (1889) described him as a pioneer in the development of more systematic and exact methods in plant breeding by introducing "measure, number and weight"; [at that time, the breeders in Europe tried to apply the Darwinian theory to plant breeding, i.e., the character of a plant could be improved by continuous selection; it began with repeated mass selection in heterogeneous local landraces of wheat, oats, barley, and peas;

often the results were disappointing]; at this time, experienced botanists such as H. TEDIN and P. BOLIN were employed at the Swedish Seed Association [as early as 1890, they started with individual selection in order to enable a better description of the range of variation available in the local varieties; during the 1890s a great number of crosses in different crops were performed and analyzed at Svalöf; it was clear that both parent plants had equal influence on the progeny; P. BOLIN pointed out from his experience with barley crosses ". . . that regarding the heredity of characters in the second and in the immediately following generations a definite regularity seems to exists: the types, which appear, represent all possible combinations of the characters from the parents, and can therefore be calculated in advance with almost mathematical accuracy." (BOLIN, 1897)]; NILSSON-EHLE emphasized as early as 1904 the possibilities of counteracting economic losses caused by plant diseases through the production of resistant varieties—he disputed the belief the cause of plant diseases could be found in insufficient access to plant nutrients; between 1908 and 1911, he published papers about the inheritance of quantitative characters in wheat and oats, and proved that these characters are inherited in the same way as qualitative characters; he strongly emphasized the need for careful planning of the cross-breeding method in the breeding work—only lines and varieties with well-known reactions and features should be chosen to achieve the expected goal; in the year 1915, he was appointed professor of botany at the University of Lund; in 1917, a personal professorship in genetics was conferred upon him, which he held from 1925, in conjunction with the directorship of the Swedish Seed Association; NILSSON-EHLE used mainly wheat and oats for his genetic studies; one of the problems he studied was the wheat "speltoid" phenomenon [however, the definite solution to the problem was presented by J. MACKEY (1954), who at that time was working at Svalöf; he concluded that the "speltoid" trait could be produced by simple deficiency, deficiency duplication, and probably also by gene mutations or chromosome substitution]; his main contribution was the discovery of the phenomenon of polymery in plants—he applied the knowledge to practical combination and population breeding of cereal and forest plants.

NOVER, I. (1915-1985): born in Kassel (Germany); she studied at the Biological and Agricultural Faculties of the Universities of Wroclaw (Poland) and Halle/Saale (Germany) from 1934 to 1938; she obtained her PhD from the University of Halle/Saale in 1941, working on mildew in wheat; in 1948, she was appointed to the Phytopathological Institute of the University of Halle/Saale, where she remained till her retirement in 1976; during her twenty-eight years' work on resistance to rusts, smuts, and mildew, she heavily contributed to resistance breeding in wheat, barley, and rye in Germany; she was one of the first plant pathologists to establish tester stock collections, evaluate wild collections, and transfer the results of basic research to breeding programs.

OEHLKERS, F. (1890-1971): a German botanist studying the mitosis of several fungi; influenced by O. RENNER of the Botanical Institute of München (Germany), he began, in 1921, genetic studies on the genus *Oenothera*; from 1934, he investigated the physiology of meiosis; during the years 1942-1943, his laboratory at the Institute of Botany of the University of Freiburg (Germany) discovered several chemical mutagens.

OKAMOTO, M. >>> SEARS, E. R.

OSMOMYSL, Y. >>> FIFE, D.

PAP, A. (1897-1992): he was born to Jewish farmer-tenant parents, with the handicap of a double dislocation of the hip; he was saved from deportation by >>> George REDEI, at that time an official in the Hungarian Ministry of Agriculture (later a professor at the University of Missouri, United States); in the late 1940s and early 1950s PAP was imprisoned; after his release from prison, he was employed at the Agricultural Research Institute in Martonvasar (Hungary); after four years he succeeded in developing the maize hybrid combination "Mv 5" and years later "Mv 1," the first inbred hybrid in Europe—in this work he also used his own hybrid lines ("Mpf 511D"), selected earlier during his life as a private farmer; his hybrids were a breakthrough—within a few years Hungary became the most significant hybrid maize seed exporter in Europe, and this position has been maintained for decades; PAP was an excellent scientist, although he wrote only a few scientific papers [however, the only Hungarian paper referred to by VAVILOV in his famous work *The Origin, Variation, Immunity and Breeding* was a publication of A. PAP]; after he left Hungary following the revolution in 1956, his followers continued the maize-breeding work he had initiated both at the Agricultural Research Institute of the Hungarian Academy of Sciences in Martonvasar and at the Cereal Research Institute in Szeged.

PAREY, P. >>> FRUWIRTH, C.

PAULI, W. >>> DELBRÜCK, M.

PAULY, E. (1905-1989): a German breeder working in Quedlinburg (Germany); she bred several varieties of *Mathiola, Callistephus, Antirrhinum, Petunia, Viola*, etc.

PAWLETT, T. L. >>> WATSON, I. A.

PEARSON, K. >>> FISHER, R. A.

PESOLA, V. A. (1892-1983): born at Turku (Finland); he studied botany and plant biology at the University of Turku; from 1918 to 1924, he worked in the Plant Breeding Station of Jaevenpaeae and later specialized in plant physiology at Jokionen (from 1924 to 1928); he became the head of the Plant Breeding Station at Jokionen in 1928; from 1930, he was the director of Plant Breeding Station of the Center for Agronomic Science at Tikkurila; he was a pioneer in opening hundreds of square kilometers for agriculture in the north, not only in his own country but also in the Scandinavian countries, Canada, and Russia; he produced more than twenty varieties for use in cold northern regions.

PIENAAR, R. de V. (1928-): he was born in Hamburg (Germany) and grew up in Johannesburg (South Africa); he graduated from the University of the Witwatersrand; after two years postgraduate study in England, Sweden, and Holland, he completed his cytological studies on *Eragrostis* at Wits and received his PhD in 1953; PIENAAR started his career as an assistant scientific officer at the Department of Agriculture in 1954, and proceeded through the ranks until he was appointed as professor of cytogenetics in 1964; upon the retirement of >>> LAUBSCHER, he was appointed as the head of the department in 1969; PIENAAR was an exceptionally hard worker, meticulous scientist, and dedicated teacher; his first appointment was related to initiating carrot-, onion-, and chinkerinchee (*Ornithologum*)-breeding programs; later, his field of interest and expertise was in wide crosses involving the genus *Triticum* and related species in order to create new allopolyploids from which gene transfers could be effected to *Triticum durum* and *Triticum aestivum* from rye, *Thinopyrum, Haynaldia, Agropyron,* or *Hordeum;* for this purpose he also created monosomic, nullisomic, and telosomic series of aneuploids in the variety "Pavon 76," and transferred the wheat crossability genes *Kr1* and *Kr2* to this cultivar; he was the first to successfully intercross the indigenous wild-growing *Thinopyrum* with bread and durum wheats; this research was well supported by postgraduate students, which led to thirteen MSc and six PhD degrees; with industry funding and technical support by H. S. ROUX, he initiated triticale-, durum wheat-, and rye-breeding programs; this led to the release of six durum wheat and five triticale cultivars; PIENAAR was well recognized internationally, traveled extensively, and was awarded by CIMMYT "in recognition of his lifelong dedication to wheat improvement through building bridges between wheat and its relatives"; in 1986, he was appointed as the third MONSANTO-SEARS Visiting Professor at the University of Missouri, Columbia; before his retirement in 1991, he initiated a haploid-breeding research project with wheat and barley, which he continued for SENSAKO Ltd. until it was taken over by Monsanto in 1998.

PISAREV, V. E. (1883-1972): a Russian agronomist, crop breeder, and geographer; the founder of Tulun Experiment Station (Russia); from 1921 he was the scientific expert of the Department of Applied Botany, and from 1925 the deputy director of VIR (Leningrad, Soviet Union).

PRIDHAM, J. T. >>> WATERHOUSE, W. L.

PUNNETT, R. C. >>> STURTEVANT, A. H.

PUSTOVOYT, V. S. (1886-1972): a Russian breeder, born in Tara-
novka, Charkovskaya guberniya (now Ukraine); he introduced the
sunflower as a crop plant, which later spread all over the world; he was
a pioneer in breeding the first sunflower cultivars; he bred thirty-four
cultivars, among them two that received worldwide fame—"Salut"
and "Peredovik."

RAATZ, W. (1864-1919): a German sugar beet breeder
in Kleinwanzleben; he developed efficient selection methods for increasing yield and sugar content by classification of E (= high yielding and normal sugar content),
N (= normal yielding and normal sugar type), Z (= high
sugar content and normal yielding), and ZZ types (=
very high sugar content and normal yielding).

RABINOVYCH, S. V. (1932-): a Russian plant breeder; in 1954, she
received a degree at Kharkov Agricultural Institute and was appointed
as an agronomist, plant breeder, and seed producer; she worked in the
Myrgorod State Variety Test Station (1954), the Forage Production
Department of Research Institute for Livestock-Farming of Ukraine
(1955-1957), and for more than forty-five years at the Institute of Plant
Production of the Ukrainian Academy of Agricultural Sciences; her
principal directions of scientific activity were the collection, study,
and conservation of plant genetic resources and their introduction
into the breeding process, particularly of wheat, rye, and triticale.

RATCHINSKY, T. (1929-1980): a Bulgarian wheat breeder working
longterm at the Institute of Wheat and Sunflower, General Toshevo/
Varna; born at Vratsa, he started his career at the Agricultural Institute in Sofia and was later appointed to the Agricultural Institute of
Knesha in 1957 and in 1963 to the Institute of Wheat and Sunflower,
General Toshevo; he bred important winter wheat varieties for Bulgaria, such as "Rusalka," "Jubilje," "Ogosta," "Rubin," "Tcharodejka,"
"Dobrudja 1," "Ludogorka," "Vega," and "Slatija," which still serve
as an important gene pool in new programs; he was the first in Bul-
garia to introduce the genes for short straw and earliness from the
Italian wheat lines into native stocks.

REDEI, G. >>> KISS, A.

RENNER, O. >>> OEHLKERS, F.

RHOADES, M. M. (1903-1991): he was born in Graham, Missouri, and spent his childhood in Downs, Kansas; he attended the University of Michigan, majoring in botany and mathematics; he was befriended by Prof. E. G. ANDERSON; ANDERSON introduced him to plant genetics; after receiving his BSc and MSc degrees at Michigan, RHOADES studied for his PhD at Cornell University under R. A. EMERSON, a maize geneticist; he was part of a brilliant group of maize cytogeneticists that included B. McCLINTOCK, C. BURN-HAM, and G. BEADLE; RHOADES interrupted his PhD work for one year to visit Caltech as a teaching fellow; at this time he worked on *Drosophila* under the guidance of >>> A. H. STURTEVANT and T. DOBZHANSKY, with occasional support from T. H. MORGAN and C. B. BRIDGES; following completion of his PhD at Cornell in 1932, he stayed on as an experimentalist in plant breeding until 1935; that year, he joined the USDA as a research geneticist and was stationed at Iowa State University until 1937; in 1937, the USDA transferred him to the Arlington Experimental Farm; at the farm, his basic cytogenetic research flourished, with considerable support from both his supervisor M. T. JENKINS and bureau chief F. D. RICHEY; he returned to academics in 1940 as associate professor at Columbia University; he was promoted to full professor in 1943 and remained at Columbia until 1948, when he was appointed professor at the University of Illinois; he spent ten years of teaching and research at Illinois, and next served as chairman and professor of the botany department at Indiana University, from 1958 to 1968; in 1968, he resigned the chairmanship and was given the rank of distinguished professor at Indiana University; during his career, RHOADES worked on a wide variety of topics in maize cytogenetics, including crossing-over and basic cytogenetic principles, cytoplasmic male sterility, centromeric misdivision, the first transposon-type mutator system, a nuclear gene (*iojap*) that affects the chloroplast genome, meiotic mutations (including *ameiotic 1*), meiotic drive by abnormal maize chromosome 10, properties of heterochromatin, and the effect of B chromosomes on heterochromatin; he served as the editor of *Maize Genetics Coopera-*

tion News Letter from 1932 to 1935 and again from 1956 to 1974; he was the editor of the journal *Genetics* from 1940 to 1948; RHOADES and DEMPSEY demonstrated that the system producing chromosomal breakage contains two components—it requires at least two B chromosomes plus a specific inbred genetic background to be effective; under these circumstances chromosomes with knobs undergo frequent chromosome breakage in the pollen; the breakage is visualized by the expression of a recessive phenotype in a homozygous recessive × homozygous dominant cross; the system became known as "high loss" due to the frequent elimination of dominant markers from knobbed chromosomes; a further value of this system was the discovery of new transposon systems; chromosome breakage in maize seems to stimulate the activation of transposons; new transposons were reported and analyzed in 1989.

RICHEY, F. D. >>> EMERSON, R. A. >>> RHOADES, M. M.

RIMPAU, W. (1842-1903): a German agronomist and breeder working in Schlanstedt; he elaborated important scientific and practical fundamentals of cereal breeding; he described for the first time the self-sterility of rye; one of the most famous commercial grain cultivars at this time was the "Schlandsteder Roggen" (Schlansted rye); it was produced by him through twenty to twenty-five years of meticulous mass selection; in 1883, he selected from a two-rowed barley a plant with branched spikes that could be bred true by constant selection; in 1877, he discovered within a red-glumed and awnless landrace of wheat three new types of spikes: (a) spikes showing awns, (b) white-glumed plants, and (c) a compactum-type of spike linked with stiff straw; he was also the first to produce, in 1888, a fertile octoploid wheat–rye hybrid, which can be taken as the birth of the triticale as a new crop plant; he knew that in 1876 the English botanist A. S. WILSON had presented stalks of two sterile wheat–rye hybrid plants in 1875 to the Edinburgh Botanical Society (United Kingdom); he mentioned also attempts on production of wheat–rye hybrids by BESTEHORN and by CARMAN (1882).

RÖMER, T. (1883-1951): a German agronomist with strong contributions to plant breeding at the University of Halle/Saale; he initiated the utilization of statistic tests in agricultural and breeding research programs on resistance and quality breeding in cereals; under his guidance more than twenty varieties of winter wheat, spring wheat, winter barley, spring barley, and oats pea were developed; he was one of the editors of the *Handbuch der Züchtung landwirtschaftlichen Kulturpflanzen* (P. PAREY Verl., Berlin), which had five volumes.

ROUX, H. S. >>> PIENAAR, R. de V.

RUDORF, W. (1891-1965): born in Rotingdorf (Westfalia, Germany), where his father was a practicing farmer; he attended elementary and a private secondary school in Werther, 1898-1907, and for four years in Bielefeld, where he passed his qualifying examination in 1913; in 1920, he enrolled at the Agricultural College of Berlin and graduated with a diploma in agricultural sciences in 1923; after moving to the University of Halle/Saale, he was recruited by >>> T. RÖMER as a candidate for a doctor's degree, and finished his thesis in 1926 with "Statistical analyses of variation in varieties and lines of oats"; from 1927 to 1929, he was an assistant to the directing manager of the large agricultural estates of WENTZEL in Teutschenthal and Salzmünde and concluded this period with an inaugural dissertation (DSc) in the field of agronomy and plant breeding entitled "Contribution to breeding for immunity against *Puccinia glumarum tritici*" at the University of Halle/Saale; in 1929, RUDORF was offered a professorship at the University of La Plata (Argentina); there he founded and established the "Instituto Fitotécnico" in Santa Catalina, and he served as its director until 1933; after a three-year period as an ordinary professor and director of the Institute of Agronomy and Plant Breeding of the University of Leipzig, RUDORF was appointed director of the Kaiser Wilhelm Institut für Züchtungsforschung at Müncheberg (Germany), following the late E. BAUR; at the same time he became professor for breeding research at the University of Berlin [in 1945, the Kaiser Wilhelm Institut was transferred to West Germany and found an almost ten-years interim shelter at a state property in Voldagsen; only in 1955 was the new location prepared for the (now) Max Planck Institute for Breeding Research in Köln-Vogelsang (Germany), and the move into modern laboratories and glasshouses, into appropriate experimental fields and administrative buildings, finally set the stage for modern breeding research];

RUDORF retired in 1961; his particular focus continued to be directed towards plant pathogen resistance and wide crosses with related species to achieve progress in wheat resistance; after the death of G. STELZNER in the last days of World War II, he took charge of the Potato Division of the Institute and engaged himself actively in experimental work not only in *Phytophthora* and virus, but also in beetle resistance; along with his own experiments, RUDORF also participated in breeding for combined *Gloeosporium* and virus resistance of *Phaseolus* beans; he conducted extensive investigations with medicinal plants such as *Datura, Digitalis,* and *Mentha,* and described induced mutants with potential breeding value, e.g. *compactum*-forms of *Festuca pratensis, unifoliata*-mutants of *Medicago sativa* and leafy, finely branched types of *Melilotus albus;* he developed a lime-tolerant alfalfa variety, a highly vigorous and productive white clover, and the low-coumarin melilot variety "Acumar"; he created synthetic rapeseed from interspecific crosses of turnip and cabbage and made considerable progress with soybean in Germany by combining earliness, yield, and other important traits of performance; RUDORF was one of the first in Germany to recognize the crucial importance of heterosis in breeding for yield characters, and he effectively supported hybrid breeding in maize by elaborating efficient breeding designs; he devoted himself to many public duties and cooperation within the German Agricultural Society and the Federal Association of Plant Breeders; he founded the most successful German study group on potato breeding and seed potato production, and continued at his Cologne institute the seminars for practical plant breeders initiated by his predecessor E. BAUR; he set out in 1958 for a several month's lecture and study trip through North and South America, and lectured as honorary professor at the University of Göttingen (1946-1955) and the University of Köln (1956-1961), there supervising thirteen PhD theses; together with >>> T. RÖMER, he edited the *Handbook of Plant Breeding,* published by P. PAREY, Berlin (1st edition in 1941-1950, Vol. 1-5; 2nd edition in 1958-1962, Vol. 1-6).

RÜMKER, K. H. von (1859-1940): a German agronomist who contributed much to scientific agronomy and to the development of new crop varieties; in 1889, RÜMKER applied the method of distinction of hereditary variation between and within certain systematic groups to breeding of races ("Sorten"); their heredity could be modified by mass selection, but new races originated from "spontaneous variations;" VRIES's mutation theory also contained this kind of hierarchy; new elementary species originate through what VRIES called "progressive mutation;" this corresponds to the creation of a new sort of "pangene."

RUSSELL, W. A. >>> BOJANOWSKI, J. S.

SAGERET, M. (1763-1851): a French naturalist and agronomist and member of the Société Royale de Centrale d'Agriculture de Paris; the botanist BROGNIART named the plant genus *Sageretia* after him; in his important hybridization experiments within the family of *Cucurbitaceae* he—for the first time in the history of plant hybridization—arranged the parental characters in an opposite scheme; in his segregation studies, the term "dominant" is clearly introduced.

SAKAMURA, T. >>> KIHARA, H.

SAUNDERS, C. E. (1867-1937): born in London; he selected and tested the famous wheat variety "Marquis," and introduced it to the Canadian west, which was the beginning of the large commercial production of high-quality bread wheat in Canada; he assisted his father in his many varied interests, such as plant hybridization and entomology; in 1903, his father, recognizing his meticulous standards and perseverance, appointed him to the Experimental Farms Service as experimentalist; SAUNDERS immediately applied scientific methods to his new task, and spent summers selecting individual heads of wheat from breeding material that previously had been selected en masse; a cross of "Hard Red Calcutta" × "Red Fife," made in 1892 by his brother A. P. SAUNDERS, resulted in a new variety, "Markham"; however, "Markham" did not produce uniform offspring, even though many plants had desirable characteristics; SAUNDERS again carefully selected individual heads from early plants that had stiff straws; he ensured that seeds from each plant were grown separately, with no mixing of strains; selection was rigorous, only the top lines being kept; he determined which lines had strong gluten by chewing a sample of kernels, and he introduced the baking of small loaves to measure volume; the best strain was named "Marquis"; in 1907, all surplus seed was sent to Indian Head (Saskatoon) for further testing; according to SAUNDERS's annual reports, the response of "Marquis" to Saskatchewan conditions was extremely good with regard to earliness and baking quality.

SAVITSKY, V. F. (1902-1965): a Russian-born plant breeder; he emigrated after the World War II to the United States; he successfully bred *monogerm* beets; he is considered to be the father of monogermity of cultivated sugar beet; his daughter, H. SAVITSKY, became a recognized beet cytogeneticist; she produced a first trisomic series of beet.

SCHELL, J. S. (1936-2003): he was one of the pioneers of plant genetic engineering technology and discoverer of the transfer mechanism of bacterial genes into plants; in 1976, he demonstrated the plasmid of *Agrobacterium tumefaciens* as a tumor-inducing vehicle in plants when he was working at the University of Gent (Belgium); after the so-called "Green Revolution" he proposed the term "Gene Revolution"; the resulting studies led to the first transgenic plants; J. G. MELCHERS proposed him consequently as new director of the Max Planck Institut für Züchtungsforschung, Köln-Vogelsang (Germany); he was a member of the New York Academy of Sciences and of the National Academy of Sciences in the United States and India, chairman of the Council of the European Molecular Biology Organization (EMBO), Heidelberg (Germany), and of the Scientific Advisory Board of the Otto Warburg Center in Rehovot (Israel), honorary member of the Academy of Arts and Sciences in Cambridge (United Kingdom), and member of the Royal Swedish Academy, Stockholm (Sweden).

SAX, K. >>> KIHARA, H.

SCHINDLER, O. (1876-1936): a German gardener and, from 1922 to 1936, director of the State Experiment and Research Station for Horticulture at Pillnitz (Germany); his interest was focused on the development of better strawberries and apple trees; in 1925, he bred the strawberry variety "Mieze Schindler," which is still and successfully grown in Germany, as well as the apple root stock "Pi 80."

SCHNELL, F. W. W. (1913-): retired professor of applied genetics and plant breeding at the University Hohenheim (Germany); after study in Germany and visits to the North Carolina State College, Raleigh (United States), he started his career at Max Planck Institute of Breeding Research in Germany (1952-1963); from 1963 to 1981, he was engaged by the University of Hohenheim as a full professor; he intensively promoted German hybrid breeding in maize and in 1953 started started the first hybrid breeding in rye, which is now successfully established in European agriculture.

SCHRIBAUX, E. >>> TSCHERMAK-SEYSENEGG, E. von

SEARS, E. R. (1910-1991): born in Bethel, Oregon; he obtained his BSc degree from the University of Oregon and graduated from Harvard University; he was appointed by the USDA as a Research Geneticist and started his long association with the University of Missouri in 1936, which continued until his retirement in 1980; due both to his theoretical and practical contributions, he became the "Father of Wheat Cytogenetics"; the discovery of a low level of female fertility in a wheat haploid and the recovery of aneuploid progeny led to the construction of a vast range of aneuploids unequalled in its versatility, practicality, and creativity in any other species known to man; with KIHARA and others, he developed the concept of chromosomes from three species contributing the genomes of bread wheat [chromosome 1 of species A codes for functions that are similar to chromosome 1 of species B, and chromosome 1 of species B codes for functions that are similar to chromosome 1 of species D; this concept of so-called "homoeology" is now fundamental to perception of all allopolyploid species]; after 1950, he changed the course of experimental manipulation of crop plants by producing interspecific chromosome additions, substitutions, and translocations; he was the first to transfer a gene for leaf rust resistance from an alien chromosome into common wheat; his joint work with the Japanese M. OKAMOTO led to a deeper understanding of how genes regulate chromosome pairing in polyploid wheat; the recognition of how homoeologous chromosomes are prohibited from pairing, thus immediately allowing a high degree of fertility in a polyploid plant, became a cornerstone of modern chromosome manipulations; in addition to his influence on molecular genetics and biotechnology, his basic contributions to wheat cytogenetics heavily influenced crop genetics in general.

SEGURA, J. >>> BEADLE, G. W.

SENGBUSCH, R. von (1898-1985): he was born in Riga (Latvia); from 1918 to 1919, he studied agronomy at the University of Halle/Saale (Germany); he received his PhD from the same university in 1924 under the supervision of >>> Prof. T. RÖMER; after several stints at Klein Wanzleben, Berlin, Müncheberg, Petkus, and Göttingen, he moved to the Max Planck Institute of Breeding Research Hamburg-Volksdorf, where he became emeritus in 1968; his main achievement was the selection of low-alkaloid lupins; his rapid seed-testing method was the basis of modern screening procedures; in 1931, he started selection for nicotine-free tobacco and cannabinol-free hemp; he contributed to the sex-specific inheritance of hemp, spinach, and asparagus; his screening method was also successfully applied to the selection of low-alkylresorcinol plants in rye; after 1945, he started strawberry breeding; in 1954, the variety "Senga Sengana" was released, showing high yields and good industrial adaptability; it became the all-round variety in the world for many years; it is still grown in many countries; SENGBUSCH represented the prototype of a scientist—educated, well-experienced, and focused.

SHARKOV, T. >>> IVANOV, I. V.

SHIRREFF, P. (1791-1876): the third son of J. SHIRREFF was born at Mungoswells (Scotland); he became a famous Scottish breeder at Haddington, where he carried out his cereal research; he was one of the best-known oats and wheat breeders of the middle decades of the nineteenth century in the United Kingdom; he published the book *Improvement of Cereals* and developed a series of varieties through the selection of useful genotypes within landraces ("Mungoswells," 1819, "Shirreff's Bearded Red," "Shirreff's Bearded White," "Pringle," 1857, and "Shirreff's Squarehead," 1882); he was one of the first to make use of hybridization between wheat varieties; his cross of "Talavera" and "Shirreff's Bearded White" resulted in the varieties

"King Richard" (1850) and "King Red Chaff White"; the oat varieties "Early Fellow," "Fine Fellow," "Long Fellow," or " Early Angus" were released around 1865 and became widespread.

SHULL, G. H. (1874-1954): an American botanist and plant geneticist who discovered the heterosis effect in maize; based on his scientific work, yields in maize production could be increased by about 50 percent; the production of the so-called "hybrid corn" is forever connected with his name; soon after the rediscovery of the Mendelian laws, he started genetic studies with Prof. DAVENPORT at Cold Spring Harbor (New York) on beans, poppy, *Melandrium, Digitalis,* and mutations in *Oenothera.*

SOMORJAI, F. (1900-1981): he retired as the director of the Cereal Research Institute, Szeged (Hungary); born in Nagykoeroes, he took a degree in agriculture in Keszthely and Budapest; in 1927, he moved to the Cereal Research Institute, Szeged; he took part in the first Hungarian experiments on rice production; he bred the "Szegedi Yellow Dent Corn," "Szegedi Angustifoliate Blue-grass," "Szegedi Wheatgrass," and "Ujszegedi Winter Oat."

STELZNER, G. >>> RUDORF, W.

SPRAGUE, G. F. >>> BOJANOWSKI, J. S.

STAKMAN, E. C. >>> WATERHOUSE, W. L.

STRAMPELLI, N. (1866-1942): a famous Italian breeder; his wheat varieties "Ardito," "Mentana," "Villa Glori," "Damiano," and "San Pastore" were distributed around the world prior to the "Green Revolution."

STRUBE, F. (1847-1897): a German agronomist and breeder working in Schlanstedt; he bred several wheat, oat, and rye varieties; he founded one of the most successful breeding companies "Fa. Friedrich Strube Saatzucht," which is still active in Germany.

STRUBE, H. (1878-1919): a German breeder at Schlanstedt; after the death of his father F. STRUBE, he became the head of the seed breeding company "Fa. Friedrich Strube Saatzucht;" in 1911, he established an experimental station in Guty (Russia) concerned with seed reproduction for the eastern seed market and began examining his own seeds regarding winter hardiness; in 1913, he founded an examination station on the outskirts of the "Lüneburger Heide" (German heathland region) to examine the cultivation on light habitat.

STUBBE, H. (1902-1989): a German geneticist who contributed much to mutation research and its application in plant breeding; he became the first director of the Kaiser Wilhelm Institut für Kulturpflanzenforschung in Vienna (Austria); that institute was transferred after World War II to Germany near Gatersleben; based on this institute, the "Institute of Crop Plant Research," Gatersleben, was established (now Institute of Plant Genetics and Crop Plant Research); one of the main ideas was the establishment of a world crop plant collection (later called a gene bank) and its utilization for crop improvement.

STURTEVANT, A. H. (1891-1970): he was the youngest of six children; his father taught college mathematics for a while but later took up farming, first in Illinois and later in southern Alabama, where the family moved when STURTEVANT was seven years old; he went to a country school and later to a public high school in Mobile; at the age of seventeen, he entered Columbia University; he was interested in genetics as the result of tabulating the pedigrees of his father's horses; he continued this interest at Columbia and also collected data on his own pedigree; he read some books on heredity, which led him to the textbook on Mendelism by R. C. PUNNETT (1907); he saw at once that Mendelism could explain some of the complex patterns of inheritance of coat colors in horses that he and others before him had observed; his brother Edgar, older by sixteen years, encouraged him to write an account of his findings and take it to T. H. MORGAN, who at that time was professor of zoology at Columbia; MORGAN encouraged STURTEVANT to publish the paper, and it was submitted to the *Biological Bulletin* in 1910; the other result of STURTEVANT's interest was the explanation of the relation between inversion sequences in different species, which he explored when he was given a desk in the famous fly room at Columbia University, where, only three months earlier, MORGAN had found the first white-eyed fly; during his Caltech period he collaborated, especially with >>> S. EMERSON, >>> T. DOBZHANSKY, and G. BEADLE; later STURTEVANT developed a keen interest in the history of science; his classic book *A History of Genetics* was published in 1965.

SUTTON, W. (1877-1916): an American geneticist; during the years 1902-1904, he worked together with T. BOVERI to demonstrate that chromosomes segregate as genes.

SWAMINATHAN, M. S. >>> ANDERSON, R. G.

TATUM, E. >>> BEADLE, G. W.

T

TAVCAR, A. (1895-1979): from 1922, he was head of the Department of Genetics and Plant Breeding of Agricultural Faculty, University of Zagreb (Croatia); he was occupied with research on maize, wheat, barley, rye, and other crops; he recognized the importance of plant resistance to certain wheat diseases (1927) and of winter hardiness (1929-1930) for stable grain production; he studied the relationship between morphological and physiological characteristics of wheat plant (1930-1934); between the two world wars, by means of individual selection from the cultivar "Sirban Prolifik," he developed the cultivar "Maksimirski Prolifik 39" and, by means of crossing and pedigree selection in segregating progeny, the cultivars "Maksimirska Brkulja 530," "Maksimirska Brkulja 540," and "Maksimirska Brkulja 24"; those varieties were grown in the western part of Croatia and Bosnia between 1930 and 1959; in the 1950s, he started pioneer work in Yugoslavia on mutation breeding of crop plants [gamma-ray-induced mutants, produced by ^{60}Co irradiation, were tested under field conditions nationwide].

TEDIN, T. >>> NILSSON-EHLE, H.

TIMIRIAZEV, K. A. >>> NAVASHIN, S. G.

TIMOFEEFF-RESSOVSKY, N. >>> GREBENSCIKOV, I. S.

TROLL, H.-J. (1906-2004): retired as professor of plant breeding from the University of Leipzig (Germany); after school in Flensburg (1922), he studied at Landwirtschaftliche Hochschule, Berlin (1924-1927); at the Kaiser Wilhelm Institut für Züchtungsforschung, Müncheberg, he received his PhD (his supervisor was Prof. E. BAUR); his name is closely associated with lupin breeding in Germany; he was able to select a white-seeded strain (No. 8) in *Lupinus luteus*—the basis of several varieties, such as "Weiko"; TROLL was the first university teacher of the author and stimulated his interest in genetics, plant breeding, and research.

TSCHERMAK-SEYSENEGG, E. von (1871-1962): born in Vienna (Austria) to a well-known Viennese family of scientists; his father, G. TSCHERMAK (1836-1927), was professor of mineralogy and his grandfather, E. FENZL (1808-1879), was professor of botany and director of the botanical garden; he received his basic education from the padres of the clerical gymnasium at Kremsmünster; in 1891, he began to study at the former Hochschule für Bodenkultur (Vienna) and simultaneously studied biology at the University of Vienna; TSCHERMAK interrupted his studies and worked for one year (1892-1893) as an agricultural volunteer in a manorial estate in Freiberg/Saxonia (Germany); friends induced him to relocate his place of study from Vienna to Halle/Saale (Germany), where he continued to study agriculture from 1893 to 1895 (BSc; PhD in 1896); his friendship with >>> K. von RÜMKER opened for him a scientific approach to plant breeding; to support his professional career, his father sponsored his training in breeding stations for vegetables and ornamentals in Stendal (1896/97) and in Quedlinburg (1897/98); he was confronted for the first time with practical breeding activities, such as mass selection in legumes, cabbages, lettuce, and ornamentals; later he acquired experience in crossing techniques, a skill that significantly determined his whole scientific life as a plant breeder; returning to Vienna in 1898, he found a position as an university assistant; simultaneously, he traveled to Gent (Belgium) and Paris (France), where he saw huge horticultural estates and legume-breeding stations; he received permission to start crossing experiments with wallflowers *(Erysimum chéiri)* and the garden pea *(Pisum sativum)* at the botanical garden in Gent; in the library he found DARWIN's book *The Cross and Self Pollination in the Kingdom of Plants,* which inspired him to search for rare pollination effects leading to xenia or neighboring pollination phenomena *(geitonogamy)* or crossing effects between individuals of the same species (heteromorphic *xenogamy*); after the experimental time in Gent, the seeds from his experiments were sent home to Vienna, while he traveled to see Dutch botanist VRIES in Amsterdam; VRIES showed him his mutants of *Oenothera lamarckiana*, but did not tell him about his recent experiments with peas; during another study tour to France (1898), he met H. L. de VILMORIN in Paris and Prof. E. SCHRIBAUX, who led the small Institute of Plant Breeding in the Agricultural College in Grignon near Paris; both of them stimulated his interest on hybridization experiments; when TSCHERMAK returned to Vienna he continued the work with the progenies of his pea crosses; he observed the phenomenon of xenia in seed pods of the same F1 plants, whereby seed color and seed shape resemble the difference in parental characters at this early stage after hybridization; these observations and the results of some backcrossing procedures, in which parental characters appeared in a 1:1 segregation scheme when the hybrids were crossed again with their parental types, formed the basis of his DSc thesis (1900); in this he demonstrated and discussed some of the results of his studies, data similar to that achieved half a century earlier by G. MENDEL (1866); soon after, TSCHERMAK became fully aware that these are fundamental principles of inheritance; to prevent his leaving for a new professorship at Brno (Technical University) and Wroclaw (University), the Viennese authorities of the "Hochschule für Bodenkultur" promoted him to the position of an assistant professor (1903) and founded in 1906 a separate Chair of Plant Breeding for him; it was the first established Chair for Plant Breeding in Europe; in 1903, at the experimental farm of the "Hochschule für Bodenkultur" in Gross-Enzersdorf, he founded the first Austrian plant-breeding station; after TSCHERMAK's travels to the United States, where he became acquainted with the steep rise of genetic science within agricultural research (TSCHERMAK and >>> RÜMKER 1910), the Prince of Liechtenstein showed interest in this prospective field of agricultural developments and in 1913 founded a new institute for plant breeding, the "G. Mendel Institute" in Lednice; there, TSCHERMAK started selection of early pea types from crosses "pois acacia" × "pois à cinq" and did nearly all his work with ornamentals; his cooperation (since 1904) with the preeminent Moravian plant breeder E. von PROSKOWETZ extended his activities particularly in the field of spring barley breeding; TSCHERMAK convinced him to change to individual selection and progeny testing; several further institutions were founded by TSCHERMAK, such as Kvasice (1904) and eighteen experimental field and plant-breeding stations in Moravia, Bohemia, Western Hungary, and Lower Austria (FEICH-TINGER 1932); the list of his varieties reflects his tremendous breeding work: rye breds—"Tschermak's Marchfelder Roggen" (1926), "Prof. Tschermak Roggen" (1927), winter wheat varieties "Weisser begrannter Marchfelder" (1909, land race reselection from Marchfeld area, later registered in 1927), "Brauner begrannter Marchfelder", "Hochschulweizen" (1928, from the Hungarian landrace Dioszeger wheat), "Zborowitzer Kolbenweizen" (reselection of Rimpau's Bastard), "Russischer Rotweizen" (Red Zborowitz wheat × White Russian wheat), "Moraviaweizen" (Edelepp × Marchfelder), "Non plus ultra" (Svalöf's Grenadier × Banatian wheat), "Non plus ultra I, II and III" (Selections of the above named cross), "Glasweizen", "Schilfweizen" (French Bon fermier × Blé gros bleu), "Excelsiorweizen" (Banater × Extra Squarehead Master); spring wheat varieties, "Znaimer" (Znaimer × Tucson (1925); two-rowed winter barley varieties, "Kirsche" (Kirsche × two-rowed Hanna × four-rowed Hanna's Riesen (1927); spring barley, "Hanna" or "Kwassitzer Original Hanna Pedigree-Gerste" (reselection of a landrace in the Hanna district, released 1908), "Hanna Kargyn" (Hanna × Turkish landrace from Kargyn, world champion in the London Exhibition 1927), "Hanna Kaisarie" (Hanna × Turkish landrace from the district Kaisarie), "Hanna × Chevalier", "Hanna × Schwarzenberggerste", "Hanna × Hannchen"; spring oat varieties, "Tschermak's Frühhafer" (1931, Svalöf's Siegeshafer × Hungarian sixty-day oat selection), "Tschermak's Gelbhafer" (Lochow's Gelbhafer × Svalöf's Goldregen), "Tschermak's Weisshafer" (Russian Ligowo × Savlöf's Goldregen); garden pea and bean varieties, "Tschermak's veredelte Victoria Erbse" (pois acacia × pois à cinq cosses), "Saxa × Buxbaum × Saxa;" garden beans, "Tschermak's fadenlose frühe Buschbohne" (Wiener Busch × weißgründige Heinrich's Riesen), "Tschermak's Feuerbohne" *(Phaseolus vulgaris × P. multiflorus);* faba bean, "Tschermak's weissblühende Ackerbohne;" oil pumpkin, "Tschermak's schalenloser Ölkürbis", as well as numerous vegetable varieties; he published more than 100 scientific papers and also innumerable articles in agricultural newspapers and garden journals to keep the agronomists and the gardeners in the mood to buy seeds or root-stockings of new varieties and to use them in a wide range of planting conditions; TSCHERMAK also carried out experimental phylogenetic research; in his different fertile intergeneric hybrids, he suspected that their existence was caused by unreduced gametes of F1 hybrids, e.g. wheat × rye or *Aegilops ovata × Triticum dicoccoides* [the latter hybrid, which he called "Aegilotricum," was the first induced and cytologically approved intergeneric goat wheat hybrid (TSCHERMAK and BLEIER 1926); it was the early beginning of synthetic polyploids for agricultural utilization]; the list of genera and species included by TSCHERMAK (1958) also demonstrates the excellent crossing techniques he developed: *Triticum × Secale, Triticum × Agropyron, Aegilops × Triticum, Haynaldia × Triticum, Agropyron × Secale,* radish *(Raphanus sativus f. radicula)* × Wild radish *(Raphanus raphanistrum), Secale montanum × S. cereale* (perennial rye), *Hordeum trifurcatum × H. compositium, Avena sativa × A. fatua, A. sativa × A. brevis, Triticum vulgare × T. spelta, T. vulgare × T. durum, T. vulgare × T. turgidum, T. vulgare × T. compositum, T. vulgare × T. dicoccum, T. vulgare × T. polonicum, T. turgidum × T. villosum;* other species, such as *Phaseolus vulgaris × P. multiflorus, Matthiola incana × M. tricuspicata, Verbascum olympicum × V. phoeniceum, Beta trigynax × B. lomatogon.*

TSUCHIYA, T. (1923-1992): a Japanese geneticist born in the Oita Prefecture; he graduated from the Gifu Agricultural College in 1943, and obtained a BSc degree from the Kyoto Imperial University in 1947, majoring in genetics; he taught biology, cytogenetics, and human genetics at two institutions in Japan before accepting a position as cytogeneticist at the prestigious Kihara Institute for Biological Research near Yokohama; the experience of working with the internationally recognized geneticist Prof. H. KIHARA was the turning point of TSUCHIYA's career; KIHARA became TSUCHIYA's lifelong mentor and friend; while working at the Kihara Institute, he completed his DSc degree at the Kyoto University in 1960; in 1963, he moved to the University of Manitoba in Winnipeg (Canada) as a postdoctoral fellow, where he worked for five years in the Department of Plant Science; in 1968, TSUCHIYA joined the Department of Agronomy at Colorado State University (United States); he filled the position vacated by D. ROBERTSON on his retirement; his research involved the genetics, cytology, cytogenetics, and cytotaxonomy of barley, sugar beet, melon species, triticale, tree species (e.g., giant sequoia), and *Alstroemeria,* the Peruvian lily; in the last years of his life he elucidated the genetic and cytogenetic effects in seeds subjected to long-term storage.

 **VALLEGA, J. (1909-1978):** born in Italy; his scientific career started in 1931 at the University of Buenos Aires (Argentina) as a plant pathologist; from 1934 to 1943, he was the head of the Phytotechnical Institute of Santa Catalina of the University of La Plata; he pioneered research on the physiological specialization of rusts on cereals and flax; in 1956, he founded *Robigo,* an Argentinean newsletter on rust diseases.

VAVILOV, N. I. (1887-1943): a Russian plant geneticist born in Moscow; he studied at the Moscow Agricultural Institute (now the Timirjasev Academy of Agriculture); in 1907, having graduated from the Agricultural Institute, he continued to work at the Department of Agriculture headed by Prof. PRYANISHNIKOV; from 1911 to 1912, he worked at the Institute of Applied Botany in St. Petersburg, which belonged to R. REGEL (1867-1920); from 1913, he visited several institutions in England, France and Germany; in 1917, he became a professor of genetics at the Agricultural Faculty of the University of Saratov; during the Civil War (1918-1920), Saratov became the scientific anchor for the Department of Applied Botany (Bureau till 1917); in 1920, VAVILOV was elected head of the Department, and soon moved to St. Petersburg together with his students and associates; in 1924, the Department was transformed into the Institute of Applied Botany (since 1930, All-Union Institute of Plant Production, VIR), which occupied the position of the central nationwide institution responsible for collecting the world's plant diversity and utilizing it for crop improvement [in 1927, those ideas were for the first time published during the International Congress of Genetics at Berlin (Germany)]; VAVILOV was recognized as the foremost plant geographer of contemporary times; to explore the major agricultural centers in Russia and abroad, VAVILOV organized and took part in over 100 collecting missions; his major foreign expeditions included ones to Iran (1916), to the United States, to Central and South America (1921, 1930, 1932), and to the Mediterranean and Ethiopia (1926-

1927); for his expedition to Afghanistan in 1924, he was awarded the N. M. PRZHEVALSKII Gold Medal of the Russian Geographic Society (from 1931 to 1940, VAVILOV was its president); these missions and his determined search for plants were based on VAVILOV's concepts in the sphere of evolutionary genetics, i.e., the "law of homologous series in variation" (1920) and the theory of the "centers of origin of cultivated plants" (1926); the first five world centers were originally delineated by him, and the last four were added by others:

1. China, including Western and Central China (Japanese millet, buckwheat, soybean, rice, common millet, foxtail millet, velvet bean, adzuki bean, turnip, Chinese yam, Chinese radish, ginger, kiwi, Chinese olive, Chinese hickories, Chinese quince, Chinese hazelnut, Oriental persimmon, loquat, litchi, apricot, peach, Japanese plum, Chinese pear, jujube, abutilon hemp, ramie, hemp, tea, camphor), 136^2
2. India, including Indo-Malaya, Indo-China, Burma, and Assam (pigeon pea, guar, millet, rice bean, urd, mung bean, cucumber, Indian mustard, black mustard, sesame, mango, tamarind, jute, sun hemp, kenaf) 2a. Indochina (Job's tears, Jacobean, rice, winged bean, mat bean, taro, winged yam, starfruit, sour and sweet oranges, pomelo, grapefruit, durian, mangosteen, mango, banana, plantain, rambutan, litchi, longan, Manila hemp, sugar palm, sugarcane), 172
3. Central Asia, including Northeastern India, Afghanistan, Turkmenistan, and Anatolia (common wheat, rye, oats, onion, garlic, carrot, spinach, walnut, apple, pistachio, apricot, pear, jujube), 42
4. West Asia, including the Transcaucasia region, Iran, Turkmenistan, and Asia Minor (emmer wheat, einkorn wheat, rye oats, barley, vetches, alfalfa, clovers, plums, pea, lentil, fig), 83
5. Near East and Mediterranean coastal and adjacent regions, including regions surrounding the Mediterranean Sea (vegetables, rape, lupins, beets, clovers, pea, lentil, flax, olive, broad bean, seradella, chickpea, barley, grass pea, lentil, lupin, pea, rye, hard wheat, einkorn, bitter vetch, broad bean, carrot, parsnip, parsley, lettuce, rapeseed, black mustard, safflower, linseed, poppy, almond, hazelnut, melon, quince, fig, walnut, apple, date palm, cherry, European plum, pomegranate, pear, grapes, celery, asparagus, artichoke), 84

6. East Africa, including Ethiopia, Eritrea, and Somalia (coffee, ricinus, sorghum millet, barley [*Hordeum vulgare convar. labile, H. vulgare convar. labile*] emmer wheat, linseed), 38
7. Southern Mexico, Central America, and Antilles (maize, *Phaseolus* beans, sweet pepper, sweet potato, cotton, sisal, cacao, cucumber, pumpkins, amaranth, sword bean, huaozontle, tepary bean, scarlet runner bean, Lima bean, common bean, maize, pepper, chili pepper, squash, tomato, vanilla, sapodilla, anona, papaya, avocado, guava, sisal, upland cotton), 49
8. South America with Bolivia, Ecuador, Peru, Chile, Bolivia, and parts of Brazil (maize, potato, tomato, cotton, peanuts, bananas, tobacco, rubber tree, quinoa, jack bean, achira, yam, sweet potato, manioc, ulluco, cashew, pineapple, Brazil nut, passion fruit, tobacco), 62
9. North America (lupins, strawberry, sunflower, American grape)

Some crops originated outside those centers, e.g. *Avena sativa* and *A. strigosa* (Europe), *Cola acuminata* (Western Africa), *Ribes grossularis* and *Rubus idaeus* (Europe); VAVILOV, the symbol of glory of the national science, was at the same time the symbol of its tragedy; as early as in the beginning of the 1930s, his scientific programs were being deprived of governmental support; in the stifling atmosphere of a totalitarian state, the institute headed by him turned into a resistance point to the pseudo-scientific concepts of T. D. LYSENKO; as a result of this controversy, VAVILOV was arrested in 1940, and his closest associates were also sacked and imprisoned; he died in the Saratov prison of dystrophia and was buried in a common prison grave.

VETTEL, F. (1894-1965): a successful German cereal breeder in Hadmersleben; some well-known wheat varieties, such as "Heines Teverson," "Heine II," "Heine IV," and "Heine VII," were bred by him.

VILMORIN, P.-L.-F. de (1816-1860): he was a famous French plant breeder; as did >>> J. GOSS in England (1822), he observed, during his experiments between 1856 and 1860, a 3:1 segregation for flower color in pea and *Lupinus hirsutus;* he was the first to introduce the principle of individual progeny testing, and in France he improved sugar beet by continuous selection.

VOGEL, O. A. (1908-1991): a native of Pilger, Nebraska; he is credited with revolutionizing wheat production in the United States; agriculture authorities asserted that his inventions in scientific research equipment indirectly contributed to expansion of world food production as much or perhaps more than his wheat-breeding did; in 1931, VOGEL began his forty-two-year career as a USDA wheat breeder stationed at Washington State University at Pullman, where he earned his doctorate degree in agronomy in 1939; his early experience in breeding for resistance to smut made him a pioneer in recognizing the importance of biological control of diseases; by 1949, he had developed the wheat variety "Brevor" by selecting partially smutted plants, giving it both specific and nonspecific forms of resistance; after 1949, in an effort to solve the lodging problem caused by long stalks, he began to experiment with a collection of semidwarf wheats utilizing the Japanese variety "Norin 10"; it became his major international contribution to providing wheat germplasm for CIMMYT's tasks and worldwide.

VORONOV, Y. N. (1874-1931): a Russian geobotanist and taxonomist; expert in subtropical plant diversity; from 1918 to 1921, he was the director of Tiflis Botanical Gardens (Georgia); in 1925, he became a research scientist of the Institute of Applied Botany of VIR (Leningrad, Soviet Union).

VRIES, H. de (1848-1935): a Dutch botanist who, in 1900, independent of but simultaneously with E. CORRENS and E. TSCHERMAK, rediscovered G. MENDEL's historic paper on principles of heredity; he became particularly known for his mutation theory; new elementary species originate by what VRIES (1901) called "progressive mutation;" this correspondes to the creation of a new sort of "pangene;" within a species there can occur "retrogressive" and "degressive" mutations, which correspond to the modification of existing pangenes; retrogression means the disappearance of characters through inactivation of pangenes; degression means the re-appearance of a character; the hereditary differences behave differently in hybridization; progressive mutations lead to what he called "unisexual" hybrids, whereas the hybrids of retrogressive as well as degressive mutations are "bisexual;" only the latter type of hybrid was the subject of MEN-

DEL's law; unisexual hybrids are constant and nonsegregating; VRIES speculative explanation for this behavior was that progressive mutations of the new type of pangenes would be unpaired in the hybrid; there is no "antagonist," i.e., a modified pangene of the same kind with which it could pair up; therefore, progressive mutation was VRIES' main interest, not Mendelism; he also thought that individual selection was the only method needed in plant breeding, and that the production of new forms through hybridization was superfluous and mass selection could only create "local races;" VRIES gave currency to the modern use of the term "mutation;" he provided major inspiration for research on the spontaneous change of hereditary factors; his book *Die Mutationstheorie (Mutation Theory)* from 1901 made a greater impact in some ways than the rediscovery of MENDEL's laws.

WALLACE, H. A. >>> BORLAUG, N. E.

WARBURG, O. (1859-1938): a famous German taxonomist and plant geographer; he explored the tropical flora of south and east Asia; later in his career he became an expert on tropical agriculture and settlement in German colonies.

WARD, M. >>> WATERHOUSE, W. L.

WATERHOUSE, W. L. (1887-1969): he was born at Maitland (Australia); he attended Chatswood Public School and Sydney Boy's High School; he was awarded the first Farrer Research Scholarship for a study on "The effects of superphosphate on the wheat yield in New South Wales"; in 1918, he was awarded a research fellowship; he went to Imperial College in London and obtained a diploma; on the return voyage to Australia via the United States, he spent some time in the Department of Plant Pathology at University of Minnesota, St Paul; there he came under the influence of E. C. STAKMAN; returning to Australia in 1921, he was appointed lecturer in plant pathology, genetics, plant breeding, and agricultural botany at the University of Sydney; in 1929, he received his DSc; when the early basic studies of the genus *Puccinia* began in 1921, there was great speculation throughout the world as to the cause of pathogenic variability; it was known that new and dangerous strains of most plant pathogens arise from time to time but their origin was often obscure [M. WARD (United Kingdom) had received some support for his proposals that pathogenicity may be increased as a result of organisms growing on a "bridging" host; the information available on wheat stem rust *(Puccinia graminis* f. sp. *tritici)* was at this stage not at all clear; it was suspected that the alternate host of this fungus played some role because it had been known for a long time that there is a connection between the barberry plant and the cereal stem rusts, barberry eradication programs in both Europe and North America being successful in reducing rust damage in cereal crops]; in Australia it was known that species of barberry were introduced as early as 1859; prior to 1921, it was widely accepted that the local wheat stem rust organism had lost the ability to infect them; this belief was dispelled by WATERHOUSE; wheat stem rust was found to comprise six strains; these strains could be separated as dikaryons by their ability or inability to attack a group of twelve different wheats; one strain, which he called race 43, was avirulent on barberry plants; others, such as 45 and 46, grew normally on it; the real advance was made in 1928; when WATERHOUSE found another rust—race 34—which, when used to infect barberry, gave rise in its progeny to two new races, 11 and 56, outstandingly different from the parent; race 34 was apparently heterozygous for the genes for virulence on the wheats "Einkorn" and "Mindum," and hence segregation had occurred; a new race, 21, was found by him on *Agropyron monticola* in 1948; about 1940, there was great satisfaction with the contribution that wheat breeders made towards a solution of rust control; "Thatcher," a hard spring wheat, was thriving in North America, and the Australian breeder MACINDOE had released "Eureka" in 1938; it was commonly accepted that stem rust was no longer to be feared; WATERHOUSE released his own resistant varieties ("Fedweb" and "Hofed"); however, he was always on guard for the fungus to attack any wheat that was currently resistant; the blow fell in 1942, when a rusty crop of "Eureka" was found; WATERHOUSE showed that the fungus had changed; it was a stepwise mutation in the fungus for virulence on plants with $Sr6$, the gene for resistance in "Eureka;" most of the changes that have been observed in the virulence involve single-gene mutations; WATERHOUSE recognized four formae speciales of the fungus, *avenae, secalis* and *lolii;* "Eureka" became so susceptible that farmers rejected it in favor of "Gabo," a wheat with a different gene for resistance *(Sr11);* WATERHOUSE's varieties "Hofed" and "Fedweb" were never widely grown; it resulted from crosses in which he attempted to transfer the rust resistance of the tetraploid wheats "Khapli Emmer" *(T. dicoccum)* and "Gaza" *(T. durum)* to hexaploids; he was attracted to the vigor of material in which "Gaza" had been backcrossed to a special accession of "Bobin" received from J. T. PRIDHAM; this accession is not unlike "Gular," a wheat of good quality and quite unlike the "Bobin" grown commercially; "Gabo," the wheat, which evolved from WATERHOUSE's early studies, was registered in 1945; by 1950, it was the standard of quality and has essentially been the basis of the prime hard wheats; the success of "Gabo" was not confined to one area of Australia; when the Rockefeller Foundation Program began to develop in Mexico, "Gabo" showed its superiority in rust resistance, earliness, and yielding ability; current Mexican lines can be attributed to the presence of this insensitivity gene; it is seldom that a single wheat variety contributes so much in so many different environments; "Gabo" and its sib, "Timstein," were both widely used in the pedigree of the Mexican wheats, e.g., "Cajeme," "Mayo," "Nainari," and others.

WATSON, I. A. (1914-1986): educated at the Universities of Sydney, Australia (BSc, 1938) and Minnesota, United States (PhD 1941); in 1938 he became an assistant lecturer in agriculture at the University of Sydney; through the T. L. PAWLETT Scholarship he worked at the Department of Plant Pathology and Plant Genetics, University of Minnesota, St. Paul from (1939-1941); after his return to Australia, he was again appointed as an assistant lecturer in agriculture University of Sydney (1941-1944), and later as a lecturer in agriculture (1944-1946), senior lecturer in agriculture 1946-1955, head of the Section of Genetics and Plant Breeding, Faculty of Agriculture (1952-1965), and associate professor in Genetics and Plant Breeding (1955-1962); he shortly returned to University of Minnesota, St. Paul (1955-1956); he became a professor of Agricultural Botany, University of Sydney (1962-1965), and the head of the Department of Agricultural Botany (1966-1976), dean of the Faculty of Agriculture (1966-1967), director of the Plant Breeding Institute, and head of the

Department of Agricultural Botany (1974-1977) before he retired; he was one of the most known Australian wheat rust researchers; the most important aspect of his research has been the development of theories and explanations for the origin of genetic variability in the wheat rust pathogen by asexual means; he also established the classification system for leaf and stem rust that are prevalent in Australia and New Zealand; based on his studies, he developed and implemented the theory of multiple gene resistance as a means of achieving lasting resistance to wheat rusts; as the junior partner of >>> W. WATERHOUSE, he has released since 1940 the varieties "Gabo" (1945), "Kendee" (1946), "Saga" (1951) and "Koda" (1955); after 1960, under his leadership, the varieties "Mendos" (1964), "Gamut" (1965), "Timgalen" (1967), "Gatcher" (1969), "Songlen" (1975), "Timson" (1975) and "Shortim" (1977) were registered; after his retirement in 1977, the varieties "Sunkota" (1981), "Suneka" (1982) and "Sunstar" (1983) were released; as a matter of interest, the origin of "Sunstar" goes back to his idea to create a purple-seeded feed wheat—however, the idea was not well received by the industry, and finally a white prime hard wheat variety was selected from that particular cross.

WATSON, J. D. (1928-): an American from Chicago; he entered the University of Chicago at age fifteen to study zoology and graduated with a BSc; in 1947, he was accepted to Indiana University in Bloomington for graduate studies; there he studied bacteriophages under >>> S. E. LURIA (1912-1991) and received his PhD in zoology in 1950; he then spent a year in Copenhagen (Denmark) as a postdoctoral researcher working on viruses; during this period, he went to a conference in Napoli (Italy), where he saw an X-ray diffraction picture of DNA by >>> M. WILKINS; after this pivotal incident, he soon arranged to move to the Cavendish Laboratory of Cambridge University (England), where he hoped to study DNA; together with F. H. C. CRICK, he was able to show that the hereditary substance of the chromosomes is deoxyribonucleic acid (DNA); he and F. CRICK shared the 1962 Nobel Prize in Physiology and Medicine with M. H. F. WILKINS.

WEISMANN, A. (1834-1914): a German biologist who developed the notion of the continuity of inherited material from generation to generation (so-called "germplasm"), thus suggesting that acquired characteristics are not inherited.

WELLS, D. G. (1917-2005): an American winter wheat breeder at the South Dakota State University from 1960 until his retirement in 1982; he received his doctoral degree from the Agronomy Department at the University of Wisconsin in 1949, and worked as a winter wheat breeder at Mississippi State College in Starkville; in 1958, he accepted an assignment through the State Department in Washington, DC, to be a member of a ten-men team to improve farm crops for Nigeria; he released eight wheat varieties ("Hume," "Winoka," "Bronze," "Gent," "Rita," "Nell," "Rose," and "Dawn") and eight germplasm lines that carried traits ranging from intermediate wheatgrass resistance to wheat streak mosaic virus.

WETTSTEIN, F. von (1895-1945): an Austrian botanist and geneticist working at the universities of Göttingen, München, and Berlin (Germany); he promoted scientific plant breeding; one of the most important biologists in Germany; he initiated and strongly promoted research on cytoplasmic inheritance in plants.

WILFAHRT, H. >>> HELLRIEGEL, H.

WILKINS, H. F. >>> WATSON, J. D.

WILLIAMSON, J. >>> BACKHOUSE, W. O.

WILSON, A. S. >>> RIMPAU, W.

WINGE, Ö. (1886-1960): a Danish botanist; he demonstrated the utility of using characteristics of chromosomes and noted different chromosome numbers in plants; he formulated the theory that polyploid series in nature arise through species hybridization and that the summation of the chromosome complements of the intercrossed species; this theory was later verified by MÜNTZING's synthesis of *Galeopsis tetrahit* >>> A. MÜNTZING.

WOOD, T. B. >>> HUNTER, H.

YATES, F. (1902-1994): one of the pioneers of 20th century statistics as a branch of applied mathematics, concerned with the collection and interpretation of quantitative data and the use of probability theory to estimate population parameters; in 1931, he was appointed as assistant statistician at Rothamsted Experimental Station (United Kingdom); in 1933, he became the head of statistics when R. A. FISHER obtained a position at University College London; at Rothamsted, he worked on the design of experiments including contributions to the theory of analysis of variance, a statistical method for making simultaneous comparisons between two or more means, and a statistical method that yields values that can be tested to determine whether a significant relationship exists between variables; after World War II, he worked on sample survey design and analysis; he became an enthusiast of electronic computers [in 1954, he obtained an "Elliott 401" for Rothamsted and contributed to the initial development of statistical computing]; his main contributions to experimental statistics were the "contingency tables" (involving small numbers and the χ^2 test, 1934), the design and analysis of factorial experiments (1937), statistical tables for biological, agricultural and medical research (with R. A. FISHER, 1938), systematic sampling (1948), selection without replacement from within strata (with probability proportional to size, together with P. M. GRUNDY, 1953), and sampling methods for censuses and surveys (1960); in 1966, he was awarded a Royal Medal from the Royal Society; he retired from Rothamsted in 1968 and became a senior research fellow at Imperial College.

ZHUKOVSKY, P. M. (1888-1975): a Russian botanist, wheat taxonomist, and pioneer of breeding research; from 1915 to 1925, he worked at Tiflis Botanical Gardens (Georgia); from 1925, he was a research scientist at the Institute of Applied Botany (Leningrad, Soviet Union); from 1951 to 1965, he was director of VIR (Leningrad), taking over the Heritage of N. I. VAVILOV.

Notes

Chapter 2

1. "Kamut" derives from the ancient Egyptian word for wheat. It is marketed as a new cereal; however, it is an ancient relative of modern durum wheat *(Triticum durum)*. It is thought to have evolved contemporarily with the free-threshing tetraploid wheats. It is also claimed that it is related to *T. turgidum,* which also includes the closely related durum wheat. It was originally identified as *T. polonicum.* Some other taxonomists believe it is *T. turanicum,* commonly called khorasan wheat. Although its true history and taxonomy are still in dispute, the great taste, texture, and nutritional qualities as well as its hypoallergenic properties are unequivocal. It is two to three times the size of common wheat with 20 to 40 percent more protein and is higher in lipids, amino acids, vitamins, and minerals.

2. In Europe, the sunflower plant was initially used as an ornamental, but the literature mentions sunflower cultivated for oil production by 1769. Most of the credit is given to PETER the Great (1672-1725), Czar of Russia. By 1830, the manufacture of sunflower oil was done on a commercial scale. The Russian Orthodox Church increased its popularity by forbidding most oil foods from being consumed during Lent; since sunflower was not on the prohibited list, it gained immediate popularity as a food. By the early 19th century, Russian farmers were growing over 800,000 ha of sunflower. During that time, two specific types had been identified: oil-type for oil production and a large variety for direct human consumption. Russian government research programs were implemented. >>> V. S. PUSTOVOIT developed a successful breeding program at Krasnodar (Russia): the oil content and yield were increased significantly. Today, the world's most prestigious scientific award for research on sunflower is known as the Pustovoit Award. By the late 19th century, Russian sunflower seed found its way into the United States. By 1880, seed companies were advertising the "Mammoth Russian" sunflower seed in catalogs. In 1970, this particular seed name was still being offered in the United States, nearly 100 years later. A likely source of this seed movement to North America may have been Russian immigrants. The first commercial use of the sunflower crop in the United States was as silage feed for poultry. In 1926, the Missouri Sunflower Growers' Association participated in what was probably the first processing of sunflower seed into oil. Canada started the first official government sunflower-breeding program in 1930. The basic plant-breeding material that was utilized came from Mennonite immigrants from Russia. Acreage spread because of increasing oil demand.

Concise Encyclopedia of Crop Improvement
© 2007 by The Haworth Press, Inc. All rights reserved.
doi:10.1300/5891_07

By 1946, Canadian farmers built a small crushing plant. Acreage spread into Minnesota and North Dakota. In 1964, the Government of Canada licensed the Russian cultivar "Peredovik." This variety produced high yields and high oil content. Acreage in the United States escalated in the late 1970s to over 2 million ha because of high European demand for sunflower oil. This European demand had been stimulated by Russian exports of sunflower oil in the previous decades, during which time concerns about cholesterol had reduced demand for animal fats. However, the Russians (and, later, Bulgarians) could no longer supply the growing demand for sunflower, and European companies looked elsewhere. Europeans imported sunflower seed that was then crushed in European mills. Western Europe continues to be a large consumer of sunflower oil today, but it depends on its own production and breeding.

3. Elevated rectangular fields are called "chinampas." Scholars of raised-field agriculture generally concur that the Mayan people of the lowlands of southern Yucatan in Mexico and parts of Belize and Guatemela developed the system, whose adoption by other agriculture-based societies then proceeded northward. To construct the chinampas system, the Aztecs removed and piled up aquatic vegetation and muck to create horticultural platforms flanked by waterways and drainage canals ("zanjas") by dredging one vertical meter of canal debris every one to four years. Trees such as willow and alder, which often grow in symbiosis with a nitrogen-fixing actinomycete (*Frankia* spp.), are planted around the islands' perimeters. The trees provide shade, increase diversity, and anchor the soil in the wet environment, thus reducing erosion. By the 1st century AD, raised-field agriculture was the modus operandi of food production in Teotihuacan, near present-day Mexico City. By the 16th century, the chinampas supplied grains, vegetables, and fruits to a quarter million Aztecs in Teotihuacan.

The chinampas system was still used in Teotihuacan in the late 19th century. Functional examples of the system persist today in Xochimilco in Mexico City and southwest Tlaxcala State, Mexico. Similar systems flourished in present-day Peru, Bolivia, and Ecuador well before COLUMBUS' arrival in the New World. Popular use of the system began to decline at the time of the Spanish Conquest. Factors that influenced this decline include salinization, population pressures, and inequitable access to technologies that affect labor use, such as plows. Recent studies are reviving interest in the chinampas system as a way to sustain food production in specific ecological conditions with minimal imported inputs.

4. The Arabs entered Spain in the 8th century AD (AD 711) and moved as far north as southern France. In 1492, the last Moors were expelled from Spain.

Chapter 3

1. In the American breeding literature, it is spelled "Oderbrucker." The "Oderbrucher" barleys are known as six-rowed, rough-awned varieties with white kernels. They are particularly prized by maltsters in the upper Mississippi Valley of the United States. They yield well and are stiff-strawed and moderately tolerant to summer heat and humidity. Oderbrucher selections were grown in the United States on

40.960 ha in 1935. Oderbrucher was originally a variety identical or similar to the "Manchuria." As with the latter variety, it consisted of a large number of strains, both blue and white. A report in the old U.S. government records of 1865 states: "This variety is grown very extensively on the low, formerly swampy lands of the valley of Oder (Germany)," although these were drained during the reign of King Frederick the Great. In German, the word for the swampland is *Bruch*, so, the barleys from there are called Oderbrucher. Evidently, by an error of translation or typing in the United States, this became Oderbrucker. An import by the federal government of the United States apparently never reached the farmers. In 1889, however, the Ontario Agricultural College at Guelph received this barley from Germany and later sent it to the Wisconsin Agricultural Experiment Station (United States). It was widely distributed by the Wisconsin Station and most of the improvement was made there. In 1908, R. A. MOORE and A. L. STONE of Madison released Wisconsin Pedigree 5 and Wisconsin Pedigree 6. Both of these were selections of Oderbrucher.

2. Dent and flint maizes differ in many traits. By about 1800, North American colonists from Europe found that hybrids between the two types could be advantageous: in particular, flints provide alleles for earlier maturity and dents for higher yield. Between 1800 and 1920, farmers in Canada and the United States, principally from flint × dent crosses, had developed over 500 open-pollinated varieties of maize. According to R. W. ALLARD (1999), the allogamous variety "Reid yellow dent" became the most widely adapted and popular maize in the United States during the late 1880s. The farmers R. REID and J. REID moved from Ohio to Illinois in 1846 and brought dent maize with them. Probably by spontaneous crosses with Indian flint maize surrounding the dent stands, a sort of a hybrid variety was established.

3. LYSENKO became famous for the discovery of *"vernalization,"* an agricultural technique that allowed winter crops to be obtained from summer planting by soaking and chilling the germinated seed for a determinate period of time. He was the first to use the term "vernalization" (in Russian, *jarovization* means inducing spring growth habit).

4. However, R. C. LEWONTIN (1977) mercilessly pointed to the misery of quantitative genetics during the 1st International Conference on Quantitative Genetics, held in 1976 at Iowa State University (United States). His major criticism was that it did not provide accurate answers to questions like: How many genes influence a trait?, What is their contribution to the phenotypic and genotypic variance?, What is their action and interaction?, How are they organized within the genome?, or How do they react in different environments?

Chapter 6

1. From the time of its discovery until 1896, teosinte was known principally to a few botanists who had preserved some dried specimens in European herbaria and bestowed upon it the Latin name, *Euchlaena mexicana*. Teosinte was placed in the genus *Euchlaena* rather than in *Zea* with maize because the structure of its ear is so profoundly different from that of maize that 19th-century botanists did not appreciate the close relationship between these plants. When the first maize × teosinte hy-

brids were discovered in the late 1800s, they were not recognized as hybrids but were considered a new and distinct species, i.e., *Zea canina*. A Mexican agronomist J. SEGURA made the first experimental maize × teosinte crosses, demonstrating that *Zea canina* was a maize × teosinte hybrid and thereby implying that maize and teosinte were much more closely related than previously thought.

2. Number of crop plants named by VAVILOV for each center of diversity.

Glossary

A comprehensive volume has been published as: *Encyclopedic Dictionary of Plant Breeding and Related Subjects,* The Haworth Press, 2003.

AFLP >>> amplified fragment length polymorphisms

allele-specific PCR (AS-PCR): Refers to amplification of specific alleles or DNA sequence variants at the same locus; specificity is achieved by designing one or both PCR primers so that they partially overlap the site of sequence difference between the amplified alleles; variants of this technique have been described under different names such as RAPD, AP-PCR, or DAF (DNA amplification finger printing).

allogamy: Cross-fertilization as opposed to autogamy.

allopolyploid: Plants with more than two sets of chromosomes that originate from two or more parents; the sets contain at least some nonhomologous chromosomes.

amplified fragment length polymorphisms (AFLPs): Polymorphic DNA fragments are amplified through PCR procedure; their differences are used for genotype identification and linkage studies.

aneuploid: A cell or organism whose nuclei possess a chromosome number that is greater or smaller by a certain number than the normal chromosome number of that species; an aneuploid results from nondisjunction of one or more pairs of homologous chromosomes.

anthropochore: A wild plant or species dispersed as a result of accidental human activity.

antisense gene: A gene construct placed in inverted orientation relative to a promoter; when it is transcribed, it produces a transcript

Concise Encyclopedia of Crop Improvement
© 2007 by The Haworth Press, Inc. All rights reserved.
doi:10.1300/5891_08

complementary to the mRNA transcribed from the normal orientation of the gene.

antisense RNA: A complementary RNA sequence that binds to a naturally occurring (sense) mRNA molecule; in this way it blocks its translation.

AS-PCR >>> allele-specific PCR

backcross breeding: A system of breeding whereby recurrent backcrosses are made to one of the parents of a hybrid, accompanied by selection for a specific character(s).

biotechnology: Any technique (e.g., recombinant DNA methods, protein engineering, cell fusion, nucleotide synthesis, biocatalysis, fermentation, cell cultures, cell manipulations) that uses living organisms or parts of them to make or modify products, to improve organisms, or to make them available for specific uses; more practically for plant breeding, applications are anther culture for haploid production, embryo/ovule culture after interspecific hybridization, genetic engineering (transformation), in vitro selection, in vitro germplasm conservation and exchange, micropropagation, cell and organ culture, somaclonal variation, somatic cell hybridization (protoplast fusion), or somatic embryogenesis.

bleeding: Exudation of the contents of the xylem stream at a cut surface because of root pressure.

block: A number of plots that offer the chance of equal growing conditions; comparisons among the entries are tested in the same block offer unbiased estimates of genetic differences.

block designs: The theory of design of experiments came into being largely through the work of R. A. FISHER and F. YATES in the early 1930s; they were motivated by questions of design of careful field experiments in agriculture; when it is desired to compare the yield of different varieties of grain, it is quite possible that there is an interaction between the environment and the variety that would alter the yields; so blocks are chosen in which the environment is fairly consistent throughout the block; in other types of experiments, in which the environment might not be a factor, blocks could be distinguished as those that receive a particular treatment; in this way, the classification of the experimental plots into blocks and varieties can be used whenever there are two factors that may influence yield.

bulk breeding: The growing of genetically diverse populations of self-pollinated crops in a bulk plot with or without mass selection, generally followed by a single-plant selection; it is a procedure for inbreeding a segregating population until the desired level of homozygosity is achieved; the seeds to grow each generation is a sample of that harvested from plants of the previous generation; it is usually used for the development of self-pollinated crops; it is an easy way to maintain populations during inbreeding; natural selection is permitted to occur, which can increase the frequency of desired genotypes compared with an unselected population; it can be used in association with mass selection with self-pollination; disadvantages are (1) plants of one generation are not all represented by progeny in the next generation, (2) genotypic frequencies and genetic variability cannot be clearly defined, and (3) natural selection may favor undesirable genotypes.

carpel: One of the female reproductive organs of the flower, comprising an ovary and usually with a terminal style tipped by the stigma.

centgener method: Developed by W. M. HAYS (United States) in the late 1880s; one of the earliest established pure-line systems of plant breeding based on 100 selected plants; the first step consists of selecting individual plants of promise, threshing these separately, and making nursery trials of their progeny; during the period of study, plots of 100 plants each are grown from each selection; besides taking notes on yield and other characters on the plot basis, the 10 better plants in each plot are selected in the field, threshed individually, and the seed of the 5 that are of greatest promise, after laboratory study, are bulked and used for the following year's centgener plot.

chromosome: A DNA-histone protein thread usually associated with RNA, occurring in the nucleus of a cell; it bears the genes, which constitute the hereditary material; each species has a constant number of chromosomes.

chromosome-mediated gene transfer: The transfer of genes within and between varieties, species, or genera by means of chromosome manipulations, such as additions, substitutions, translocations, or directed recombinations, utilizing specific crossing techniques, cell manipulations, or micromanipulation of chromosomes; more specifically, the use of isolated metaphase chromosomes as a vehicle for the transfer of genes between cultured cells.

chromosome mutation: Any structural change involving the gain, loss, or translocation of chromosome parts; it can arise spontaneously or be induced experimentally by physical or chemical mutagens; the basic types of chromosome mutations are deletions (deficiencies), duplications, inversions, and translocations.

clonal propagation: Vegetative (asexual) propagation from a single cell or plant.

clonal selection: Choosing the best clones from a clonal testing (e.g., in potato or forest trees).

clonal test: Evaluation of genotypes by comparing clones in a plantation.

clone: A group of genetically identical cells or individuals, derived from a common ancestor by asexual mitotic division; in molecular biology, a population of genetically identical organisms or cells; sometimes it refers to cells containing a recombinant DNA molecule or to the recombinant DNA molecules themselves; in horticulture or agriculture, a group of individuals originally taken from a single specimen and maintained in cultivation by vegetative propagation.

clone variety: Refers to a crop variety that consists of individuals deriving from a single clonal genotype (monogenotypic), e.g., in potato, cassava, or sweet potato.

cloning: The process of vegetatively propagating a certain crop and/or plant; in molecular genetics, the cloning of DNA molecules from prokaryotic or eukaryotic sources as part of a bacterial plasmid or phage replicon; usually cells are separated mechanically from outflowing medium and then added back to the culture.

combination breeding: A breeding method that utilizes the genetic diversity of individuals or varieties in order to create and to select new phenotypes on the basis of genetic recombination of useful characters of parental material.

composite cross-population: A population generated by hybridizing more than two varieties and/or lines of normally self-fertilizing plants and then propagating successive generations of the segregating population in bulk in specific environments so that natural selection is the principal force acting to produce genetic change.

convergence breeding (convergent improvement): A breeding method involving the reciprocal addition to each of two inbred lines of the dominant favorable genes lacking in one line and present in the other; backcrossing and selection are performed in parallel, each of the original lines serving as the recurrent parent in one series.

crop plant: A plant species expressly cultivated for use; the majority of crops can be classified as (1) root and tuber crops (potato, yams), (2) cereals (wheat, oats, barley, rye, rice, maize), (3) oil and protein crops (rapeseed, pulses), (4) sugar crops (sugar beet, sugarcane), (5) fiber crops (cotton, jute), or (6) forage crops (grasses, legumes); agronomic crops can be classified as (a) green manure crops, (b) cover crops, (c) silage crops, or (d) companion crops; about 2 percent of the 250,000 higher plant species are used in agriculture, horticulture, etc. (about 1,700-2,000 species); economically, the most important families are the legumes and the grasses, which account for more than a quarter of the total species; they are followed by *Rosaceae, Compositae, Euphorbiaceae, Labiatae,* and *Solanaceae.*

cytoplasmic male sterility (CMS): Pollen abortion due to cytoplasmic factors that are maternally transmitted but act only in the absence of pollen-restoring genes; this type of sterility can also be transmitted by grafting.

design of experiments: A procedure that can be used interactively to form experimental designs of various types: (1) orthogonal hierarchical designs, such as randomized blocks, split-plots, or split-split-plots; (2) factorial designs (with blocking)—these have several treatment factors and a single blocking factor (giving strata for blocks and plots within blocks; (3) fractional factorial designs (with blocking)—there are several treatment factors, but the design does not contain every treatment combination and so some interactions are aliased;

(4) lattice designs for a single treatment factor with number of levels that is the square of some integer k; (5) lattice squares, which are similar to lattices except that the blocking structure with the replicates has rows crossed with columns; (6) Latin squares, available for any number of treatments—where feasible, more than one orthogonal treatment factor can be generated to form Graeco- Latin squares; (7) Latin squares balanced for carry-over effects, (8) semi-Latin squares—that is, $n \times n$ Latin squares whose individual plots are split into k subplots to cater for a treatment factor with $n \times k$ levels; (9) alpha designs, which have a single treatment factor but no constraint on the number of levels; (10) cyclic designs, which have a single blocking factor that defines blocks that are too small to contain every treatment; (11) balanced-incomplete-block designs, where the experimental units are grouped into blocks such that every pair of treatments occurs in an equal number of blocks; (12) neighbor-balanced designs, which allow an adjustments to be made for the effect that a treatment may have on adjacent plots; and (13) central composite designs used to study multidimensional response surfaces

dihaploid: A haploid cell or individual containing two haploid chromosome sets—not to be confused with doubled haploid.

dimorphism: The occurrence of two forms of individuals within one population or other taxa.

dioecious: Possessing male and female flowers or other reproductive organs on separate, unisexual, individual plants (e.g., in hemp, hops, asparagus, and spinach).

DNA microarrays: A powerful, versatile, and economical molecular technique for screening of genetic aberrations; high-density gene sequences are printed onto glass slides; fluorophore-labeled genomic or complimentary DNA (cDNA) is hybridized to slides with fixed signature patterns and resolved using computer-driven fluorescent imaging.

dominance: The quality of one of a pair of alleles that completely suppresses the expression of the other member of the pair when both are present; the degree of dominance is expressed by the ratio of additive genetic variance to total phenotypic variance; if the ratio equals 1 then the trait shows complete dominance, if the ratio is greater than 1 then the trait shows overdominance, and if the ratio is less than 1 then the trait shows incomplete dominance.

explant: An excised fragment of a tissue or an organ used to initiate an in vitro culture.

first-division restitution (FDR): The result of an abnormal orientation of the spindles right before meiotic anaphase II; nonsister chromatids end up in the same nucleolus; it was found in several crop species (e.g., potato and rye).

gamete: A specialized haploid cell whose nucleus, and often cytoplasm, fuses with that of another gamete in the process of fertilization, thus forming a diploid zygote.

gametocidal gene: A gene encoding a product that destroys cells that divide to produce the gametes.

geitonogamy: When neighboring flowers of the same plant can achieve pollination; as opposed to xenogamy.

grafting: The joining together of parts of plants by holding cut surfaces in position until a union of living cells forms; the united parts will continue their growth as one plant; this technique is used regularly to asexually propagate fruit trees and other woody plants that do not readily root from cuttings; while nearly all of the plants that are commonly grafted can be produced from seed, the variability in the plants produced is so great that it is more practical to clone these plants by grafting; grafting was a horticultural art for several thousand years, and in the past three centuries the potential was realized for using selected rootstocks to affect growth and performance of scions of plants; it is used to accelerate a breeding program, to test for compatibility, and to determine if symptoms are virus caused; four common graft methods are bridge, bud, cleft, and whip.

half-sib progeny selection >>> method of overstored seeds (*syn* remnant seed procedure)

hemizygous: An individual, generally diploid, having a given gene and/or allele present once (e.g., in monosomics or haploids).

heritability: A measure of the degree to which a phenotype is genetically influenced and can be modified by selection; it is represented by the symbol h^2; this equals V_{gen} / V_{phe}, where V_{gen} is the variance due to genes with additive effects and V_{phe} is the phenotypic variance; the two types of heritability are (1) broad-sense heritability, $h_b^2 = V_{gen}/V_{phe}$, and (2) narrow-sense heritability, $h_n^2 = V_{add}/V_{dom}$.

heterosis: The increased vigor of growth, survival, and fertility of hybrids as compared with the two homozygotes; it usually results from crosses between two genetically different, highly inbred lines; it is always associated with increased heterozygosity; in breeding, the three types of heterosis are (1) F1 yielding more than the mean of the parents, (2) F1 yielding more than the best-yielding parents, and (3) F1 yielding more than the best-yielding variety; for the genetic basis of heterosis, two major hypotheses—dominance hypothesis and over-dominance hypothesis—have received the most attention.

heterostyly: A polymorphism among flowers that ensures cross-fertilization through pollination by visiting insects; flowers have anthers and styles of different length.

Latin rectangle: A field design similar to the Latin square but differentiated by the number of replications, which is not equal to the number of variants; the number of replications may be a third, a quarter, or a fifth of the number of variants.

Latin square: In general, a set of symbols arranged in a checkerboard in such a fashion that no symbol appears twice in any row or column; it is used for subdividing plots of land for agricultural and breeding experiments so that treatments can be tested even though the field's soil conditions might vary in an unknown fashion in different areas; it requires that the field be subdivided by a grid into subplots and that the differing treatments be performed at consecutive intervals to plants from different subplots.

linkage map: An abstract map of chromosomal loci, based on experimentally determined recombinant frequencies, that shows the relative positions of the known genes or other (DNA) markers on the chromosomes of a particular species; the more frequently two given characters recombine, the further apart are the genes that determine them.

marker-aided selection, marker-assisted selection (MAS): Indirect selection exploiting the association between the qualitative variation in a trait (isoenzymes or DNA markers) and the quantitative variation in another trait; it is a strategy permitting plant selection at the juvenile stage from early generations; the essential requirements for MAS in plant breeding are (1) marker(s) should cosegregate or be closely linked (<1 cM) with the desired trait, (2) an efficient means of screening large populations for molecular markers should be available, and (3) the screening technique should have high reproducibility across laboratories.

mass selection (positive or negative): A form of breeding in which individual plants are selected on the basis of their individual advantages and the next generation propagated from the aggregate of their seeds; the easiest method is to select and multiply together those individuals from a mixture of phenotypes, which correspond to the breeding aim (positive mass selection); it is still applied in cross-pollinating of vegetable species (e.g., carrots, radishes, beetroots) in order to improve the uniformity; when all undesired off-types are rouged in grown crop population and the remaining individuals are propagated further, the method is termed negative mass selection; negative mass selection is no longer an adequate breeding method for highly advanced varieties and is now usually applied in multiplication of established varieties.

method of overstored seeds: In pedigree breeding of allogamous crop plants (e.g., in rye), an effective method of regulating cross-fertilization; usually, a greater number of individual plants is harvested from a genotypic mixture of a certain population; their progenies are sown as A families in smaller plots while half of the seed of all elite plants is retained in reserve; those A families meeting all the requirements are not directly multiplied, but the remaining seed of the corresponding elite plants is sown in the following year; it enters a so-called A′ family trial; in this way the economically valuable traits of the A family can be definitely evaluated after maturity.

micropropagation: Vegetative propagation by application of tissue culture that is usually conducted in growth chambers; several stages are distinguished: stage I, establishment of small fragments of stock plant(s) in tissue culture; stage II, multiplication of propagules—the most common method is through stimulation of branching and subsequent division of shoot clumps on smaller explants, which are then placed on a fresh medium; stage III, preparation of propagules for transfer to normal growing conditions through rooting or elongation of shoots; stage IV, establishment of stage II or III propagules in normal growing conditions, usually in soil or potting mix in a greenhouse.

molecular marker: Particular DNA sequences and/or segments that are closely linked to a gene locus and/or a morphological or other characters of a plant and that can be detected and visualized by molecular techniques; roughly, three groups of markers can be classified: (1) hybridization-based DNA markers, such as restriction fragment length polymorphisms (RFLPs) and oligonucleotide fingerprinting; (2) PCR-based DNA markers, such as random amplified polymorphic DNAs (RAPDs), which can also be converted into sequence-characterized amplified regions (SCARs), simple sequence repeats (SSRs) or microsatellites, sequence-tagged sites (STS), amplified fragment length polymorphisms (AFLPs), inter-simple sequence repeat amplification (ISA), cleaved amplified polymorphic sequences (CAPs), or amplicon length polymorphisms (ALPs); and (3) DNA chip and sequencing-based DNA markers, such as single-nucleotide polymorphisms (SNPs); the advantages of molecular markers are (a) the precision (<1 cM) by which molecular markers can be used to map a site on a genome is often better than phenotypic markers, (b) linkage can be easily established, (c) the markers need not be genes—they can be any piece of DNA that detects a small number of bands and for which there is polymorphism in the population, which makes them selectively neutral, and (d) for RFLPs, a detected polymorphism need not fall inside the region of the probe itself.

monocarpic: Bearing one fruit.

monogerm(ous): A fruit (e.g., sugar beet) containing only one ovule; in contrast to a multigerm fruit, which is an aggregate fruit containing several ovule units.

mutagen: An agent that increases the mutation rate within an organism or cell—for example: X-rays, gamma-rays, neutrons, or chemicals (base analogues, such as 5-bromo uracil, 5-bromo deoxyuridine, 2-amino purine, 8-ethoxy caffeine, 1,3,7,9-tetramethyl-uric acid, and maleic hydrazide); antibiotics, such as azaserine, mitomycin C, streptomycin, streptonigrin, and actinomycin D; alkylating agents, such as sulfur mustards (ethyl-2-chloroethyl sulfide), nitrogen mustards (2-chloroethyl-dimethyl amine), epoxides (ethylene oxide), ethyleneimines, sulfates, sulfonates, diazoalkanes, and nitroso compounds (N-ethyl-N-nitroso-urea); azide (sodium azide); hydroxylamine; nitrous acid; and acridines (hydrocyclic dyes), such as acridine orange.

mutation: A change in the structure or amount of the genetic material of an organism; in cytogenetics, a gene or a chromosome set that has undergone a structural change.

mutation breeding: To experimentally introduce or remove a character from a cell or plant by exposure to mutagenic agents followed by screening for the desired attribute; it also refers to several techniques involving induced mutations that were utilized to introduce desirable genes into the plants.

nondisjunction: The failure of separation of paired chromosomes at metaphase, resulting in one daughter cell receiving both while the other daughter cell receives none of the chromosomes in question; it can occur both in meiosis and mitosis.

off-type: An individual differing from the population norm in morphological or other traits; the term also includes escapes and contaminants (e.g., seeds that do not conform to the characteristics of a variety, uncontrolled self-pollination during production of hybrid seed, segregates from plants).

pangenesis: Recapitulation of certain ancestral traits during embryogenesis.

parallel mutation: A mutation that causes similar phenotypes but in different species.

parthenogenesis: The development of an individual from an egg without fertilization.

parthenote: A cell or individual resulting from parthenogenesis.

pedigree breeding: A system of breeding in which individual plants are selected in the segregating generations from a cross on the basis of their desirability judged individually and on the basis of a pedigree record; the advantages are: (1) if selection is effective then, inferior genotypes may be discarded before lines further evaluated, (2) selection in generation involves a different environment, which provides a good genetic variability, and (3) the genetic relationship of lines is estimated and can be used to maximize genetic variability.

pendulous: Drooping or hanging downward.

plant breeding: The science and practice of altering the genetic pattern of plants in order to increase their value; increased crop yield is the primary aim; advantages of the hybrids and new varieties developed include adaptation to new agricultural areas, greater resistance to disease and insects, greater yield of useful parts, better nutritional content of edible parts, and greater physiological efficiency; other goals are adaptation of crops to modern production techniques, such as mechanical harvesting and improvement in the market quality of the product; traditionally, plant breeders have made genetic changes in crops by using various crossing and selection methods; attempts have also been made to introduce favorable mutations; increasingly, desirable traits and/or genes are being introduced into cultivated plants via genetic engineering.

pleiotropic: An allele or gene that affects several traits at the same time.

point mutation: A mutation that can be mapped to one specific locus; it is caused by the substitution of one nucleotide for another; it may also be caused by deletion and inversion.

polycross: Open pollination of a group of genotypes (generally selected) in isolation from other compatible genotypes in such a way as to promote random mating *inter se;* it is a widely used procedure for intercrossing parents by natural hybridization.

polycross test: A progeny test to assess "general combining ability" from crosses among selected parents; identities can be maintained only for the seed parents; a mixture of pollen is artificially applied to each female parent.

polyembryony: The condition in which an ovule has more than one embryo, as in certain grasses or cereals; in the past, the phenomenon was used for haploid selection among the embryos, which often show different ploidy levels.

polymerase chain reaction (PCR) technique: A technique for continuous amplification of DNA and/or DNA fragments in vitro; the DNA sequence must be known so that oligonucleotides can be synthesized that are complementary to the extremes of the fragment to be amplified; heat-stable DNA polymerase (e.g., from *Thermus aquaticus*) is used for DNA synthesis.

polysomy: The reduplication of some but not all of the chromosomes of a set beyond the normal diploid number.

progeny test: A test of the value of a genotype based on the performance of its offspring produced in some definite system of mating.

protandry: The maturation of anthers before carpels (e.g., in sugar beet, sunflower, and carrot).

protogyny: A condition in which the female parts develop first (e.g., in rapeseed).

pyramidal: Triangular in outline; shaped like a pyramid.

random amplified polymorphic DNA (RAPD) technique: A comparative study (among individuals, populations, or species) of the DNA fragment length produced in controlled DNA synthesis reactions started with short sequences of DNA (primers); as a genetic mapping methodology, it is based on the fact that specific DNA sequences (polymorphic DNA) are repeated (i.e., appear in sequence) with a gene of interest and so the polymorphic DNA sequences are linked to that specific gene; their linked presence serves to facilitate genetic mapping within a genome.

RAPD >>> random amplified polymorphic DNA technique

rDNA-ITS >>> ribosomal DNA internal transcribed spacers

recombination: The process whereby new combinations of parental characters may arise in the progeny, caused by exchange of genetic material of different parental lines.

restriction fragment length polymorphism (RFLP): A comparative study (in individuals, populations, or species) of the DNA fragment lengths produced by particular restriction enzymes; by using a DNA hybridization technique, restriction fragments can be identified if they are complementary to a specific DNA probe; each mutation that produces or eliminates a restriction site in a homologous region leads to a change of length of the restriction fragment, which must be detected in order to infer genomic relationships; RFLPs represent an important tool in detecting variability—they are free of secondary effects due to pleiotropic action and are frequently associated with the segregation of alleles affecting morphophysiological traits.

RFLP >>> restriction fragment length polymorphism

ribosomal DNA internal transcribed spacers (rDNA-ITS): A multigene family with nuclear copies in eukaryotes that are arranged in tandem arrays in nucleolar organizer regions (NORs), generally at more than one chromosomal location; each unit within a single array consists of genes coding for small (18S) and large (28S) rRNA subunits; the 5.8 S nuclear rDNA gene lies embedded between these genes but are separated by the two internal transcribed spacers ITS1 and ITS2.

scion: A portion of a shoot or a bud on one plant that is grafted onto a stock of another.

secondary crop: A crop that originated as a weed of a primary crop (e.g., rye); in agronomy, a crop grown after a primary crop.

selection: The process determining the relative share allotted individuals of different genotypes in the propagation of a population; natural selection occurs if zygotic genotypes differ with regard to fitness.

sequence-tagged-site (STS) marker: A unique (single-copy) DNA sequence used as a mapping landmark on a chromosome.

simple-sequence repeat (SSR) DNA marker technique: A genetic mapping technique that utilizes the fact that microsatellite sequences repeat (appear repeatedly in sequence within the DNA molecule) in a manner enabling them to be used as markers.

single-nucleotide polymorphisms (SNPs): Variations (in individual nucleotides) that occur within DNA at the rate of approximately one in every 1,300 bp in most organisms; SNPs usually occur in the same genomic location in different individuals.

single-polymorphic amplified test (SPLAT): If sequence-tagged-site (STS) do not reveal polymorphism, it is usually converted into SPLAT; individual STS products from different genotypes are themselves sequenced; any difference revealed can be sequences from nuclear ribosomal DNA; they can be exploited in the production of internal primers; in general, single-polymorphic amplified sequences are dominant markers.

single-seed descent (SSD): Derivation of plants by a selection procedure in which F2 plants and their progeny are advanced by single seeds until genetic purity is achieved; single-seed descent methods (single-seed, single-hill, multiple-seed) are easy ways to maintain populations during inbreeding.

single-strand conformational polymorphism (SSCP): A feature that relies on secondary and tertiary structural differences between denatured and rapidly cooled amplified DNA fragments that differ slightly in their DNA sequences; different SSCP alleles are resolved on nondenaturing acrylamide gels, usually at low temperatures; the ability to resolve alleles depends on the conditions of electrophoresis, and this requires DNA sequence data.

site-directed mutagenesis (SDM): The process of introducing specific base pair mutations into a gene; a technique that can be used to make a protein that differs slightly in its structure from the protein that is normally produced; single mutation is caused by hybridizing the region in a codon to be mutated with a short, synthetic oligonucleotide, causing the codon to code for a different specific amino acid in the protein gene product; site-directed mutagenesis holds the potential to create modified (engineered) proteins that have desirable properties not currently available in the proteins produced by the plant.

SNP >>> single-nucleotide polymorphisms

somaclonal variation: Somatic (vegetative nonsexual) plant cells can be propagated in vitro in an appropriate nutrient medium; depending on the composition and conditions, the cells may proliferate in an undifferentiated (disorganized) pattern to form a callus or in a differentiated (organized) manner to form a plant with a shoot and root; the cells that multiply by division of the parent somatic cells are called somaclones and, theoretically, should be genetically identical with the parent; in fact, in vitro cell culture of somatic cells—whether from a leaf, stem, root, shoot, or cotyledon—frequently generates cells that are significantly different, genetically, from the parent; during culture, the DNA breaks up and is reassembled in different sequences, giving rise to plants whose identifiable characteristics differ from those of the parent; such progeny are called somaclonal variants and provide a useful source of genetic variation.

SPLAT >>> single-polymorphic amplified test

SSCP >>> single-strand conformational polymorphism

SSR >>> simple sequence repeat

standard deviation: A measure of the variability in a population of items; that is, a set of n measurements $x_1, x_2, x_3, ..., x_n$ is equal to the positive square root of the variance of the measurements.

standard error: A measure of variation of a population of means.

STS >>> sequence-tagged-site marker

synthetic (variety): A variety produced by crossing *inter se* a number of genotypes selected for good combining ability in all possible hybrid combinations, with subsequent maintenance of the variety by open pollination; usually the first generation of a synthetic variety is obtained by a polycross involving a certain number of components with a good general combining ability; the components are maintained by identical reproduction, either by vegetative propagation (clones) or by continued sib mating (inbred populations).

Taq (DNA) polymerase: A DNA-dependent RNA polymerase from phage T7, which recognizes a very specific promoter sequence; it is used in many expression vectors.

topcross: A cross between a selection, line, clone, etc. and a common pollen parent, which may be a variety, inbred line, single cross, etc.; the common pollen parent is called the topcross or tester parent and is normally used to test the "general combining ability."

Union for Protection of New Varieties of Plants: Union pour la Protection des Obtentions Vegetales (UPOV), an intergovernmental organization with headquarters in Geneva (Switzerland); it is based on the International Convention for the Protection of New Varieties of Plants, as revised since its signature in Paris on December 2, 1961; on April 16, 1993, the Union consisted of 23 member states; the objective of the convention is the protection of new varieties of plants by an intellectual property right.

unisexual: A flower that possesses either stamens or carpels but not both (i.e., a plant possessing only male or female flowers).

variable number tandem repeat (VNTR): Genetic markers that consist of DNA segments that are duplicated end to end; the number of copies present at a locus can vary, giving rise to a large number of alleles; these genetic markers are commonly used for DNA fingerprinting.

vernalization: The treatment of germinating seeds with low temperatures to induce flowering at a particular preferred time.

VNTR >>> variable number tandem repeat

xenia: A situation in which the genotype of the pollen influences the developing embryo of the maternal tissue (endosperm) of the fruit to produce an observable effect on the seed.

xenogamy (cross-pollination): Intercrossing between flowers of different individuals; as opposed to geitonogamy.

Bibliography

Ahloowalia, B. S., Maluszynski, M. and Nichterlein, K. 2004. Global Impact of Mutation-Derived Varieties. *Euphytica* 135: 187-204.

Åkerman, A. and MacKey, J. 1948. The Breeding of Self-Fertilized Plants by Crossing. Some Experiments During 60 Years Breeding at the Swedish Seed Association. Svalöf 1886-1946, History and Present Problems. *C. Bloms Boktryckeri, Lund.*

Allard, R. W. 1999. *Principles of Plant Breeding.* J. Wiley & Sons, New York.

Ammerman, A. J. and Cavalli-Sforza, L. L. 1984. *The Neolithic Transition and the Genetics of Populations in Europe.* Princeton Univ. Press, Princeton.

Anderson, E. N. 1988. *The Food of China.* Yale Univ. Press, New Haven, Connecticut.

Anderson, W. T. 1996. *Evolution Isn't What It Used To Be.* W. H. Freeman & Co., New York.

Anonymous. 1998. State of the World's Plant Genetic Resources for Food and Agriculture. Report of UN Food and Agricultural Organization (FAO), Rome.

Arakawa, T., Chong, K. X. and Langridge, W. H. R. 1998. Efficacy of a Food Plant-Based Oral Cholera Toxin B Subunit Vaccine. *Nature Biotechnol.* 16: 292-297.

Artsaenko, O., Kettig, B., Fiedler, U., Conrad, U. and Düring, K. 1998. Potato Tubers as a Biofactory for Recombinant Antibodies. *Molecular Breed.* 4: 313-319.

Asimov, I. 1967. *The Egyptians.* Houghton-Mifflin, Boston.

Barton, J. H. and Berger, P. 2001. Patenting Agriculture. *Issues in Science and Technology* 17: 43-50.

Barton, K. A., Whiteley, H. R. and Yang, K. S. 1987. *Bacillus thuringiensis* Delta-Endotoxin Expressed in Transgenic *Nicotiana tabacum* Provides Resistance to Lepidopteran Insects. *Plant Physiol.* 85: 1103-1109.

Bartos, P. and Bares, I. 1971. Leaf and Stem Rust Resistance of Hexaploid Wheat Cultivars Salzmünder Bartweizen and Weique. *Euphytica* 20: 435-440.

Bateson, W. 1894. *Materials for the Study of Variation Treated with Special Regard to Discontinuity in the Origin of Species.* Mac Milland & Co., London.

Bateson, W. 1895. The Origin of Cultivated *Cineraria. Nature* 51: 605-607.

Bateson, W. 1899. Hybridization and Cross-Breeding as a Method of Scientific Investigation. *J. Roy. Hort. Sco.* 24: 59-66.

Bateson, W. 1907. The Progress of Genetics Since the Rediscovery of Mendel's Papers. *Prog. Rei Bot.* 368-418.

Bateson, W. 1909. *Mendel's Principles of Heredity.* Cambridge Univ. Press, New York.

Concise Encyclopedia of Crop Improvement
© 2007 by The Haworth Press, Inc. All rights reserved.
doi:10.1300/5891_09

Baur, E. 1922. *Einführung in Die Experimentelle Vererbungslehre.* Verl. Gebr. Bornträger, Berlin.

Beadle, G. W. 1970. Alfred Henry Sturtevant (1891-1970). *Yearbook Amer. Philos. Soc.:* 166-171.

Beal, W. J. 1887. *Grasses of North America for Farmers and Students.* Thorp and Gedfrey, Lansing.

Beaven, E. S. 1935. Discussion on Dr. Neyman's Paper. *J. Roy. Statist. Soc.,* Suppl. 2: 159-161.

Bernardo, R. 1998. A Model for Marker-Assisted Selection Among Single Crosses with Multiple Genetic Markers. *Theor. Appl. Genet.* 97: 473-478.

Berrall, J. S. 1966. *The Garden: An Illustrated History.* Viking Press, New York.

Bevan, M. W., Flavell, R. B. and Chilton, M. D. 1983. A Chimeric Antibiotic Resistance Gene As a Selectable Marker for Plant Cell Transformation. *Nature* 304: 184-187.

Bhalla, P. L. and Singh, M. B. 2004. Engineered Allergens for Immunotherapy. *Current Opinion in Allergy and Clinical Immunology* 4: 569-573.

Biffin, R. H. 1905. Mendel's Laws of Inheritance and Wheat Breeding. *J. Agric. Sci.* 1: 4-48.

Birchler, J. A., Auger, D. L. and Riddle, N. C. 2003. In Search of the Molecular Basis of Heterosis. *Plant Cell* 15: 2236-2239.

Blakeslee, A. F. 1921. Types of Mutations and Their Possible Significance in Evolution. *Amer. Natural.* 55: 254-267.

Blakeslee, A. F. and Avery, A. G. 1937. Methods of Inducing Doubling of Chromosomes in Plants. *J. Heredity* 28: 393-411.

Blakeslee, A. F., Belling, J. and Farnham, M. E. 1920. Chromosomal Doublication and Mendelian Phenotype in *Datura* Mutants. *Science* 52: 388-390.

Blohmeyer, E. 1877. Vom Versuchsfeld des Landw. *Inst. Leipzig. Frühlings Landw. Ztg.* 402.

Bolin, P. 1897. Nagra Iakttagelser Ofver Vissa Karaktarers Olika Nedarfningsformaga Vid Hybridisiering Hos Korn. *Sveriges Utsadesforen. Tidskr.* 7: 137-147.

Bolley, H. L. 1901. Flax Wilt and Flaxsick Soil. *North Dakota Agric. Exp. Sta. Bull.* 50: 27-60.

Borlaug, N. 2000. Ending World Hunger. The Promise of Biotechnology and the Threat of Antiscience Zealotry. *Plant Physiol.* 124: 487-490.

Borlaug, N. E. 1997. Feeding a World of 10 Billion People: The Miracle Ahead. *Plant Tissue Culture Biotechnology* 3: 119-127.

Bos, I. and Caligari, P. 1995. *Selection Methods in Plant Breeding.* Chapman & Hall, New York.

Botstein, D., White, R., Skolnick, M. and Davis, R. W. 1980. Multiple Forms of Enzymes: Construction of Genetic Linkage Map in Human, Using Restriction Fragment Length Polymorphism. *Amer. J. Hum. Genet.* 32: 314-331.

Brandt, P. 2004. *Transgene Pflanzen.* Birkhäuser Verl., Basel, Boston, Berlin.

Brar, D. S. and Khush, G. S. 1994. *Cell and Tissue Culture for Plant Improvement.* In: A. S. Basra (ed.), *Mechanism of Plant Growth and Improved Productivity. Modern Approaches.* Marcel Dekker Inc., New York, Basel, Hong Kong.

Brettell, R. I. S., Banks, P. M., Cauderon, Y., Chen, X., Cheng, Z. M., Larkin, P. J. and Waterhouse, P. M. 1988. A Single Wheatgrass Chromosome Reduces the Concentration of Barley Yellow Dwarf Virus in Wheat. *Ann. Appl. Biol.* 113: 599-603.

Bridges, C. B. 1916. Non-Disjunction As a Proof of the Chromosome Theory of Heredity. *Genetics* 1: 587-596.

Brim, C. A. 1966. A Modified Pedigree Method of Selection in Soybeans. *Crop Sci.* 6: 20-23.

Brown, A. H. D., Munday, J. and Oram R. N. 1989. Use of Isozyme-Marked Segments from Wild Barley (*Hordeum spontaneum*) in Barley Breeding. *Plant Breed.* 100: 280-288.

Brown, D. C. W. and Thorpe, T. A. 1995. Crop Improvement Through Tissue Culture. *World J. Microbiol. Biotechnol.* 11: 409-415.

Bruce, A. B. 1910. The Mendelian Theory of Heredity and the Augmentation of Vigor. *Science* 32: 627-628.

Brush, S. B. 2004. *Farmers' Bounty: Locating Crop Diversity in the Contemporary World.* Yale Univ. Press, New Haven, Connecticut.

Bytebier, B., Deboeck, F., De Greve, H., van Montagu, M. and Hernalsteens, J. P. 1987. T-DNA Organization in Tumor Cultures and Transgenic Plants of Mono-cotyledon *Asparagus officinalis. Proc. Natl. Acad. Sci. USA* 84: 5345-5349.

Caldwell, B. E., Schillinger, J. A., Barton, J. H., Qualset, C. O., Duvick, D. N. and Barnes, R. F. 1989. Intellectual Property Rights Associated with Plants. *Amer. Soc. Agron.* 52: 1-206.

Capelle, W. 1949. Theophrast über Pflanzenentartung. *Museum Helveticum* (B. Schwabe Verl., Basel) 6: 57-84.

Caplin, S. M. and Steward. F.C. 1948. Effect of Coconut Milk on the Growth of the Explants from Carrot Root. *Science* 108: 655-657.

Carleton, M. A. 1907. Development and Proper Status of Agronomy. *Proc. Amer. Soc. Agron.* 1: 17-24.

Carlson, P. S., Smith, H. H. and Dearing, R. D. 1972. Parasexual Interspecific Plant Hybridization. *Proc. Natl. Acad. Sci. USA* 69: 2292-2294.

Carter, G. F. 1977. A Hypothesis Suggesting a Single Origin of Agriculture. In C. A. Reed (ed.), *The Origins of Agriculture* pp. 89-133. Mouton Publ., The Hague.

Castle, W. E. 1903. Laws of Heredity of Galton and Mendel and Some Laws Govering Race Improvement by Selection. *Proc. Amer. Acad. Arts Sci.* 39: 323-342.

Castle, W. E. 1946. Genes, Which Divide Species or Produce Hybrid Vigor. *Proc. Natl. Acad. Sci. USA* 32: 145-149.

Chase, S. S. 1963. Analytic Breeding in *Solanum tuberosum* L.—A Scheme Utiliz-ing Parthenotes and Other Diploid Stocks. *Can. J. Genet. Cytol.* 5: 359-363.

Chee, M., Yang, R., Hubbell, E., Berno, A., Huang, X. C., Stern, D., Winkler, J., Lockhart, D. J., Morris, M. S. and Fodor, S. P. A. 1996. Accessing Genetic Infor-mation with High-Density DNA Arrays. *Science* 274: 610-614.

Chen, F. and Hayes, P. M. 1989. A Comparison of *Hordeum bulbosum*-Mediated Haploid Production Efficiency in Barley Using In Vitro Floret and Tiller Cul-ture. *Theor. Appl. Genet.* 77: 701-704.

Chilton, M. D., Drummond, M. H., Merio, D. J., Sciaky, D., Montoya, A. L., Gordon, M. P. and Nester, E. W. 1977. Stable Incorporation of Plasmid DNA into Higher Plant Cells: The Molecular Basis of Crown Gall Tumorigenesis. *Cell* 11: 263-271.

Clarke, J. A. and Bayles, B. B. 1935. Classification of Wheat Varieties Grown in the United States. *USDA Bull.* 459.

Clarke, J. G. G. and Hutchinson, J. 1977. *The Early History of Agriculture.* British Academy, Oxford Univ. Press.

Coe, E. H. 1959. A Line of Maize with High Haploid Frequency. *Amer. Natural* 91: 381-385.

Coffman, W. R. 1982. Gallery of Cereal Breeders—Neal F. Jensen. *Cer. Res. Comm.* 10: 247-257.

Collard, B. C. Y., Jahufer, M. Z. Z., Brouwer, J. B. and Pang, E. C. K. 2005. An Introduction to Markers, Quantitative Trait Loci (QTL) Mapping and Marker-Assisted Selection for Crop Improvement: The Basic Concepts. *Euphytica* 142: 169-196.

Collins, G. N. and Kempton, J. H. 1911. Inheritance of Waxy Endosperm in Hybrids of Chinese Maize. *Proc. 4th Int. Genet. Congr., Paris*: 347-356.

Comstock R. 1978. Quantitative Genetics in Maize Breeding. In D. Walden D. (ed.) *Maize breeding and genetics* pp. 191-206. J. Wiley & Sons, New York.

Comstock, R. 1996. *Quantitative Genetics with Special Reference to Plant and Animal Breeding.* Iowa State Univ. Press, Ames.

Coons, G. H. 1953. Breeding for Resistance to Diseases. *USDA Yearbook Agricult.*: 174-192.

Correns, C. 1900. Gregor Mendels Regel über das Verhalten der Nachkommenschaft der Bastarde. *Ber. Deutsch. Bot. Gesell.* 18: 158-168.

Correns, S. 1899. Untersuchungen über Xenien bei *Zea mays. Ber. Deutsch. Bot. Ges.* 17: 410-417.

Cox, D. R. 1951. Some Systematic Experimental Designs. *Biometrika* 38: 312-323.

Crossway, A., Oakes, J. V., Irvine, J. M., Ward, B., Knauf, V. C. and Shewmaker, C. K. 1986. Integration of Foreign DNA Following Microinjection of Tobacco Mesophyll Protoplasts. *Mol. Gen. Genet.* 202: 179-195.

Curtis, B. C., Rajaram, S. and Macpherson, H. G. 2002. Bread Wheat—Improvement and Production. FAO of United Nations, Rome.

Darby, W. J., Ghalioungui, P. and Grivetti, L. 1977. *Food: The Gift of Osiris.* 2 vol. Academic Press.

Darwin, C. 1883. *The Variation of Animals and Plants Under Domestication.* 2 vols. 2nd ed. New York, D. Appleton & Co (1st ed. London, John Murray, 1868).

Davenport, C. B. 1908. Degeneration, Albinism and Inbreeding. *Science* 28: 454-455.

Decaisne, J. 1855. Le jardin fruitier du muséum, ou monographie des arbres fruitiers cultivés dans cet établissement. Examen critique de la doctrine de van Mons. *Journal de la Société Impériale et Centrale d'Horticulture* 1: 218-240.

Delaunay, L. N. 1931. Resultate eines dreijährigen Röntgenversuchs mit Weizen. *Züchter* 3: 129-137.

Dittrich, M. 1959. Getreideumwandlung und Artproblem. *G. Fischer Verl.*, Jena.

Dramer, I. N. 1981. *History Begins at Sumer.* Univ. of Pennsylvania Press, Philadelphia.

Driscoll, C. J. and Anderson, L. M. 1967. Cytogenetic Studies of Transec–A Wheat-Rye Translocation Line. *Can. J. Genet. Cytol.* 9: 375-380.

Duchnesne, M. 1766. *Histoire naturelle des Fraisiers.* Didot le jeune et C. J. Panckoucke, Paris.

Duke, J. A. 1983. *Medicinal Plants of the Bible.* Trado-Medic Books, Owerri, New York, London.

Dunn, L. C. 1965. *A Short History of Genetics.* McGraw-Hill, New York.

Durant, W. 1954. *The Story of Civilization. Part I. Our Oriental Heritage.* Simon & Schuster, New York.

Dustin, A. P. 1934. Colchicine in Agriculture, Medicine, Biology. *Bull. Acad. Roy. Med., Belgium* 14: 487-502.

Duvick, D. N. 1997. What is Yield? In G. O. Edmeades et al. (ed.) *Proc. Symp. Developing Drought- and Low N-Tolerant Maize* pp. 332-335. CIMMYT, El Batan, Mexico. 332-335.

Duvick, D. N. and Cassman, K. 1999. Post-Green Revolution Trends in Yield Potential of Temperate Maize in the North-Central United States. *Crop Sci.* 39:1622-1630.

East, E. M. 1908. Inbreeding in Corn. *Rept. Connecticut Agric. Exp. Stat. for 1907:* 419-428.

East, E. M. and Jones, D. F. 1920. Genetic Studies on the Protein Content of Maize. *Genetics* 5: 543-610.

Ehdaie, B., Whitkus, R. W. and Waines, J. G. 2003. Root Biomass, Water-Use Efficiency and Performance of Wheat-Rye Translocations of Chromosome 1 and 2 in Spring Bread Wheat "Pavon." *Crop Sci.* 43: 710-717.

Elliot, F. C. 1958. *Plant Breeding and Cytogenetics.* McGraw-Hill Book Co., Inc. New York.

Endo, T. R. and Gill, B. S. 1996. The Deletion Stocks of Common Wheat. *J. Heredity* 87: 295-307.

Evans, D. A. and Sharp, W. R. 1983. Single Gene Mutations in Tomato Plants Regenerated from Tissue Culture. *Science* 221: 949-951.

Farooq, S. and Azam, F. 2002. Molecular Markers in Plant Breeding: Concepts and Characterization. *Pak. J. Biol. Sci.* 5: 1135-1140.

Faust, M. and Timon, B. 1995. Origin and Dissemination of Peach. *Hort. Rev.* 17: 331-379.

Feichtinger, E. K. 1932. Die Entwicklung und die praktische Tätigkeit der Lehrkanzel für Pflanzenzüchtung an der Hochschule für Bodenkultur in Wien und der Pflanzenzuchtstation in Groß-Enzersdorf. *Zeitschr. Pflanzenzüchtung* 17: 1-7.

Fischer, R. and Schillberg, S. 2004. *Molecular Farming.* Wiley–VCH Verl., Weinheim, Germany.

Fisher, R. A. 1918. The Correlation Between Relatives on the Sup Position of Mendelian Inheritance. *Transact. Roy. Soc., Edinburgh* 52: 399-433.

Fisher, R. A. 1925. *Statistical Methods for Research Worker.* Oliver & Boyd, London.

Fletcher, H. R. 1969. *Story of the Royal Horticulture Society (1804-1869).* Oxford Univ. Publ. for RHS., Oxford.

Flores, H. E. and Filner, P. 1985. Metabolic Relationships of Putrescine, ABA, and Alkaloids in Cell and Root Cultures of *Solanaceae*. In K.-H. Neumann, W. Barz, and E. Reinhard (eds.), *Primary and Secondary Metabolism of Plant Cell Cultures*. Springer Verl., Berlin, Stuttgart.

Focke, W. O. 1881. *Die Pflanzen-Mischlinge, ein Beitrag zur Biologie der Gewächse*. Verl. Gebr. Bornträger, Berlin.

Fortin, M. 1999. *Syria: Land of Civilizations*. Ed. Musée de la civilisation de Québec, Québec.

Fraley, R.T., Rogers, S. B., Horsch, R. B., Sanders, P. R., Flick, J. S., Adams, S., Bittner M. L. Brand, L. A., Fink, C. L., Fry, J. S., Galluppi, G. R., Goldberg, S. B., Hoffmann, N. L. and Woo, S. C. 1983. Expression of Bacterial Genes in Plant Cells. *Proc. Natl. Acad. Sci. USA* 80: 4803-4807.

Framond, A. J., Bevan, M. W., Barton, K. A., Flavell, R. and Chilton M.D. 1983. Mini-Ti plasmid and a chimeric gene construct: New approaches to plant gene vector construction. *Adv. Gene Techn.* 20: 159-170.

Frandsen, H. N. 1940. Some Breeding Experiments with Timothy. *Imp. Agric. Bur. Joint Pub.* 3: 80-92.

Fromm, M., Taylor, L. P. and Walbot, V. 1985. Expression of Genes Transferred into Monocot and Dicot Plant Cells by Electroporation. *Nature* 319: 791-793.

Fussell, B.1992. *The Story of Corn*. Alfred Knopf, Inc, New York.

Gager, C. S. 1908. Effects of Rays of Radium on Plants. *Mem. New York Botan. Garden* 4: 1-278.

Galil, J. 1968. An Ancient Technique for Ripening Sycomore Fruit in East-Mediterranean Countries. *Econ. Bot.* 22: 178-191.

Galton, F. 1889. *Natural Inheritance*. MacMillan & Co., London.

Gärtner, C. F. von 1849. *Versuche und Beobachtungen über die Bastarderzeugung im Pflanzenreich*. K. F. Hering & Co., Stuttgart.

Gautheret, R. J. 1955. The Nutrition of Plant Tissue Cultures. *Ann. Lev. Plant Physiol.* 6: 433-484.

Gengenbach, B. G. and Green, C. E. 1975. Selection of T-Cytoplasm Maize Callus Cultures Resistant to *Helminthosporium maydis* race T pathotoxin. *Crop. Sci.* 15: 645-649.

Goldschmidt, R. 1911. *Einführung in die Vererbungswissenschaft*. W. Engelmann, Leipzig, Germany.

Goodspeed, T. H. and Clausen, R. E. 1928. Interspecific Hybridization in *Nicotiana*. VIII. The *sylvestris-tomentosa-tabacum* Hybrid Triangle and Its Bearing on the Origin of *Tabacum*. *Univ. Calif. Publ. Botany* 11: 245-256.

Goodspeed, T. H. and Olson, A. R. 1928. The Production of Variation in *Nicotiana* Species by X-ray Treatment of Sex Cells. *Proc. Natl. Acad. Sci. USA* 14: 66-69.

Goor, A. and Nurock, M. 1968. *The Fruits of the Holy Land*. Israel Univ. Press, Jerusalem.

Goss, J. 1822. On the Variation in the Colour of Peas Occasioned by Cross-Impregnation. *Transact. Hort. Soc. London* 5: 234-237.

Goulden, C. H. 1939. Problems in Plant Selection. *Proc. 7th Int. Genet. Congr.* Edinburgh, 132-133.

Grace, E. S. 1997. *Biotechnology Unzipped: Promises and Realities.* J. Henry Press, Washington, DC.

Guha, S. and Maheshwari, S. C. 1964. In Vitro Production of Embryos from Anthers of *Datura. Nature* 204: 497.

Gustafsson, Å. 1947. Mutations in Agricultural Plants. *Hereditas* 33: 1-100.

Haberlandt, G. 1902. Kulturversuche mit isolierten Planzenzellen. S. B. Weisen, *Naturw. Wien* 111: 69-92.

Hacking, I. 1965. *Logic of Statistical Inference.* Cambridge Univ. Press, UK.

Hagen, von V. W. 1957. *Ancient Sun Kingdoms of America.* The World Publishing Company, Ohio.

Hallo, W. W. 2000. *The Context of Scripture: Monumental Inscriptions from the Biblical World.* Vol. 3, Brill, Leiden, Boston, Köln.

Hammer, K. 1984. Das Domestikationssyndrom. *Kulturpflanze* 32: 11-34.

Hänsel, H. 1962. Die Bedeutung Tschermaks für Züchtungsforschung und praktische Pflanzenzüchtung. *Verhand. zoolog.-botan. Gesell. in Wien* 102: 13-17.

Hardy, G. H. 1908. Mendelian Proportions in a Mixed Population. *Science* 28: 49-50.

Hareuveni, N. 1984. *Tree and Shrub in Our Biblical Heritage.* Neot Kedumin Ltd., Kiryat, Ono, Israel.

Harlan, H. V. and Martini, M. L. 1938. The Effect of Natural Selection in a Mixture of Barley Varieties. *J. Agric. Res.* 57: 189-199.

Harlan, H. V. and Pope, M. N. 1922. The Germination of Barley Seeds Harvested at Different Stages of Growth. *J. Heredity* 13: 72-75.

Harlan, J. R. 1992. *Crops and Man.* 2nd ed. ASA, Madison.

Harland, S. C. 1936. Haploids in Polyembryonic Seeds of Sea Island cotton. *J. Heredity* 27: 229-231.

Harrington, J. B. 1937. The Mass-Pedigree Method in the Hybridization Improvement of Cereal. *J. Amer. Soc. Agron.* 34: 270-274.

Harris, J. A. 1912. A Simple Test of the Goodness of Fit of Mendelian Ratios. *Amer. Natural.* 46: 741-745.

Harris, J. A. 1915. Studies on Soil Heterogeneity. *Amer. Natural.* 49: 430-454.

Hayes, H. K. and Johnson, I. J. 1939. The Breeding of Improved Selfed Lines of Corn. *J. Amer. Soc. Agron.* 31: 710-724.

Hayes, H. K., Rinke, E. H. and Tsiang, Y. S. 1946. Experimental Study of Convergent Improvement and Backcrossing in Corn. *Minnesota. Agric. Exp. Sta. Techn. Bull:* 172.

Hayes, H. K. and Stakman, E. C. 1921. Wheat Stem Rust from the Standpoint of Plant Breeding. *Proc. 2nd Ann. Meet. Western Canadian Soc. Agron.* 22-35.

Hazarika, M. H. and Rees, H. 1967. Genotypical Control of Chromosome Behaviour in Rye. X. Chromosome Pairing and Fertility in Autotetraploids. *Heredity* 22, 317-322.

Helbaek, H. 1959. Domestication of Food Plants in the Old World. *Science* 130: 365-372.

Heller, K. J. 2003. *Genetically Engineered Food.* Verl. Wiley–VCH GmbH & Co. KGaA, Weinheim, Germany.

Herrera-Estrella, L., Depicker, A. Montagu, M. van and Schell, J. 1983. Expression of Chimaeric Genes Transferred into Plant Cells Using a Ti-Plasmid-Derived Vector. *Nature* 303: 209-213.

Hiatt, A. C., Cafferkey, R. and Bowdish, K. 1989. Production of Antibodies in Transgenic Plants. *Nature* 342: 76-78.

Hodgson, R. W. 2002. *Horticultural Varieties of Citrus. Citrus Industry,* Vol. 1, Univ. of California.

Hoffmann, H. 1869. *Untersuchungen zur Bestimmung des Werthes von Species und Varietät: ein Beitrag zur Kritik der Darwinschen Hypothese.* J. Ricker'sche Buchhandlung Gießen, Germany.

Hoffmann, W. 1951. Ergebnisse der Mutationszüchtung. *Vortr. Pflanzenzüchtung,* Bonn, 36-53.

Hogben, L. 1957. *The Contemporary Crisis or the Uncertainties of Uncertain Inference. Statistical Theory,* W. W. Norton & Co., Inc., Reprinted in *The Significance Test Controversy*—A. Reader, D. E. Morrison and R. E. Henkel, (eds.), 1970, Aldine Publ. Co., UK.

Hougas, R. W. and Peloquin, S. J. 1958. The Potential of Potato Haploids in Breeding and Genetic Research. *Amer. Potato J.* 35: 701-707.

Hutchinson, J., Clark, J. G. G., Jope, E. M. and Riley, R. 1977. *The Early History of Agriculture.* Univ. Press, Oxford.

Huxley, A. 1978. *An Illustrated History of Gardening.* The Lyons Press, New York.

Hyams, E. 1971. *A History of Gardens and Gardening.* Praeger, New York.

Iltis, H. 1932. *Life of Mendel.* George Allen & Unwin, London.

Itallie van Emden, W. 1940. Interview with Beijerinck. Martinus Willem Beijerinck, His Life and His Work. *App. J. Gravenshage,* Netherlands.

Jain, S. M., Shahin, E. A. and Sun, S. 1988. Interspecific Protoplast Fusion for the Transfer of Atrazine Resistance from *Solanum nigrum* to Tomato (*Lycopersicon esculentum* L.). *Plant Cell Tissue Organ Culture* 12: 189-192.

James, C. 2003. *Global Status of Commercialized Transgenic Crops. ISAAA Briefs,* Ithaca, New York.

Janick, J. 2001. Asian Crops in North America. *Hortic.Technology* 11: 510-513.

Janick, J. 2002. Ancient Egyptian Agriculture and the Origins of Horticulture (pp. 23-39). In S. Sansavini and J. Janick (eds.) *Proc. of the Intern. Symp. on Mediterranean Horticulture Issues and Prospects. Acta. Hortic.* 582: 55-59.

Janick, J. 2004. Long-Term Selection in Maize. *Plant Breed. Rev.* 24: 1-384.

Jenkins, M. T. and Brunson, A. M. 1932. Methods of Testing Inbred Lines of Maize in Crossbred Combinations. *J. Amer. Soc. Agron.* 24: 523-530.

Jennings, H. S. 1918. Disproof of a Certain Type of Theories of Crossing Over Between Chromosomes. *Amer. Natural.* 52: 247-261.

Jensen, N. F. 1965. Multiline Superiority in Cereals. *Crop Sci.* 5: 566-568.

Jessen, K. F. W. 1864. *Botanik der Gegenwart und Vorzeit in culturhistorischer Entwicklung.* Republ. in 1948 by The Chronica Botanica, Waltham Mass.

Johannsen, W. 1903. *Über die Erblichkeit in Populationen und in reinen Linien.* G. Fischer Verl., Jena.

Johannsen, W. 1926. *Elemente der exakten Erblichkeitslehre.* G. Fischer Verl., Jena.

Johnson, N. L. 1948. Alternative Systems in the Analysis of Variance. *Biometrika* 35: 80-87.

Johnson, S. W. 1891. *How Crops Grow.* Orange Judd & Co., New York.

Jones, D. F. 1917. Dominance of Linked Factors as a Means of Accounting for Heterosis. *Genetics* 2: 466-479.

Jones, D. F. 1920. Selection of Self-Fertilized Lines as the Basis of Corn Improvement. *J. Amer. Soc. Agron.* 12: 77-100.

Jones, D. F. 1945. Heterosis Resulting from Degenerate Changes. *Genetics* 30: 527-542.

Jones, L. R. and Gilman, J. C. 1915. The Control of Cabbage Yellows through Disease Resistance. *Wisconsin Agric. Exp. Sta. Res. Bull,* 38.

Karp, A. 1991. On the Current Understanding of Somaclonal Variation. *Oxford Surveys of Plant Molecular and Cell Biology* 7: 1-58.

Karp, A. 1995. Somaclonal Variation as a Tool for Crop Improvement. *Euphytica* 85: 295-302.

Karp, A., Kresovich, S., Bhat, K. V., Ayand, W. G. and Hodgkin, T. 1997. Molecular Tools in Plant Genetics Resources Conservation: A Guide to the Technologies. *IPGRI Tech. Bull,* 2.

Kasha, K. J. and K. N. Kao. 1970. High-Frequency Haploid Production in Barley (*Hordeum vulgare* L.). *Nature* 225: 874-876.

Katayama, Y. 1935. Karyological Comparison of Haploid Plants from Octoploid Aegilotriticum and Diploid Wheat. *Jap. J. Bot.* 7: 349-380.

Kattermann, G. 1937. Zur Cytologie halmbehaarter Stämme aus Weizenroggen-bastardierung. *Züchter* 9: 196-199.

Keeble, F. C. and Pellew, C. 1910. The Mode of Inheritance of Stature and of Time of Flowering in Peas (*Pisum sativum*). *J. Genetics* 1: 47-56.

Kellerman, W. A. and Swingle, W. T. 1889. Report on the Loose Smuts of Cereals. *Kansas Agric. Exp. Sta. Ann. Rpt.* 213-288.

Kempton, R. and Fox, P. 1997. *Statistical Methods for Plant Variety Evaluation.* Chapman & Hall, New York.

Kihara, Y. 1951. Triploid Watermelons. *Proc. Amer. Soc. Hortic. Sci.* 58: 217-230.

Kimber, G., 1983, Gallery of Cereal Workers—Ernest Robert Sears. *Cer. Res. Comm.* 11: 175-178.

King, I. P., Reader, S. M. and Miller, T. E. 1988. Exploitation of the "Cuckoo" Chromosome (4S) of *Aegilops sharonensis* for Eliminating Segregation for Height in Semi-dwarf *Rht2* Bread Wheat Cultivars. *Proc. 7th Int. Wheat Genet. Symp.,* Cambridge, 373-341.

Klein, T. M., Wolf, E. D., Wu, R. and Sanford, J. C. 1987. High Velocity Micro-Projectiles for Delivering Nucleic Acids into Living Cells. *Nature* 327: 70-73.

Knott, D. R. 1971. The Transfer of Genes for Disease Resistance from Alien Species to Wheat by Induced Translocations. *Mut. Breed. Disease Res.,* IAEA, Vienna: 67-77.

Koivu, K. 2004. Novel Sprouting Technology for Recombinant Protein Production. *Plant Science* 167: 173-182.

Konzak, C. F., Randolph, L. F. and Jensen, N. F. 1951. Embryo Culture of Barley Species Hybrids. Cytological Studies of *Hordeum sativum* × *Hordeum bulbosum*. *J. Heredity* 42: 124-134.

Kostoff, D. 1929. An Androgenic *Nicotiana* Haploid. *Zeitschr. Zellforsch.* 9: 640-642.

Krens, F. A., Molendijk, L., Wullems, G. J. and Schilperoort, R. A. 1982. In Vitro Transformation of Plant Protoplasts with Ti-plasmid DNA. *Nature* 296: 72-74.

Kuckuck, H. 1959. Neuere Arbeiten zur Entstehung des hexaploiden Kulturweizens. *Zeitschr. Pflanzenzüchtung* 41: 205-226.

Kynast, R. G., Riera-Lizarazu, O., Vales, M. I., Okagaki, R. J., Maquieira, S. B., Chen, G., Ananiev, E. V., Odland, W. E., Russell, C. D., Stec, A. O., Livingston, S. M., Zaia, H. A., Rines, H. W. and Phillips, R. L. 2001. A Complete Set of Maize Individual Chromosome Additions to the Oat Genome. *Plant Physiol.* 125: 1216-1227.

Kyozuka, J., Hayashi, Y. and Shimamoto, K. 1987. High Frequency Plant Regeneration from Rice Protoplasts by Novel Nurse Culture Methods. *Mol. Gen. Genet.* 206: 408-413.

Larkin, P. J. and Scowcroft, W. R. 1981. Somaclonal Variation: A Novel Source of variability from Cell Cultures for Plant Improvement. *Theor. Appl. Genet.* 60: 197-214.

Larue, C. D. 1936. The Growth of Plant Embryos in Culture. *Torrey Bot. Club Bull.* 69: 332-341.

Laurie, D. A. and Bennett, M. D. 1988. The Production of Haploid Plants from Wheat × Maize Crosses. *Theor. Appl. Genet.* 76: 393-397.

Laxton, T. 1866. Observations on the Variations Effected by Crossing in the Color and Character of the Seed of Peas. *Rep. Int. Hortic. Exhib. Bot. Congr.* 156-158.

Laxton, T. 1872. Notes on Some Changes and Variations in the Offspring of Cross-Fertilized Peas. *J. Roy. Hortic. Soc., London* 3: 10-14.

Lee, J. M. and Oda, M. 2003. Grafting of Herbaceous Vegetable and Ornamental Crops. *Hortic. Rev.* 28: 61-124.

Lee, M. 1998. Genome Projects and Gene Pools: New Germplasm for Plant Breeding? *Proc. Natl. Acad. Sci. USA* 95: 2001-2004.

Lehmann, E. 1916. Aus der Frühzeit der pflanzlichen Bastardierungskunde. *Arch. Gesch. Naturw. Techn.* 78-81.

Lein, A. 1973. Introgression of a Rye Chromosome to Wheat Strains by Georg Riebesel–Salzmünde After 1926. *Proc. EUCARPIA Symp. Triticale,* Leningrad, 158-167.

Lemieux, B., Aharoni, A. and Schena, M. 1998. Overview of DNA Chip Technology. *Mol. Breed.* 4: 277-289.

Leonard, J. N. 1973. *First Farmers.* Time Life Books, New York.

Levetin, E. and McMahon, K. 1996. *Plants and Society.* W. C. Brown, Dubuque.

Li, J. and Yuan, L. 2000. Hybrid Rice: Genetics, Breeding, and Seed Production. *Plant Breed. Rev.* 17: 15-120.

Lindström, E. W. 1929. A Haploid Mutant in Tomato. *J. Heredity* 30: 23-30.

Liu, B. L. 1997. *Statistical Genomics: Linkage, Mapping, and QTL Analysis.* CRC Press, Boca Raton.

Logan, J. 1739. *Experimenta et Meletamata de Plantarum Generations.* Leyden, Holland.

Lonnquist, J. H. and McGill, D. P. 1956. Performance of Corn Synthetics in Advanced Generations of Synthesis and after Two Cycles of Recurrent Selection. *Agron. J.* 48: 249-253.

Lörz, H. and Wenzel, G. 2005. *Molecular Marker Systems in Plant Breeding and Crop Improvment.* Springer Verl., Berlin, Heidelberg, New York.

Löve, A. 1953. Chromosome Number Reports XCIII. *Taxon* 35: 897-899.

Maan, N. N. and Gordon, J. 1988. Compendium of Alloplasmic Lines and Amphidiploids in the *Triticeae. Proc. 7th Int. Wheat Genet. Symp.*, Cambridge, 1325-1369.

MacKey, J. 1954. Neutron and X-ray Experiments in Wheat and a Revision of the Speltoid Problem. *Hereditas* 40: 65-180

Marchant, J. 1719. Observations sur la Nature des Plantes. *Hist. Acad. Roy. Sci.*, Amsterdam, Netherlands.

Martini, R. 1871. *Der mehrblütige Roggen, Verl.* A. W. Kafemann, Danzig, Deutschland.

Martini, S. 1961. Giorgio Gallesio, Pomologist and Precursor of Gregor Mendel. *Fruit Varieties and Horticultural Digest*, East Lancing.

Marton L., Wullems, G. J. and Molendijk, L. 1979. In Vitro Transformation of Cultured Cells from *Nicotiana tabacum* by *Agrobacterium tumefaciens. Nature* 277: 129-131.

Matsuura, H. 1933. *A Bibliographical Monograph on Plant Genetics (Genic Analysis). 1900-1929*, 2nd ed. Hokkaido Imp. Univ., Sapporo, Japan.

McClintock, B. 1929. Chromosome Morphology in *Zea mays. Science* 69: 629.

McFadden, E. S. 1930. A Successful Transfer of Emmer Characters to Vulgare Wheat. *Agron. J.* 22: 1020-1034.

Medina-Filho, H. P. 1980. Linkage of *Aps-1, Mi* and Other Markers on Chromosome 6. *Rep. Tomato Genet. Coop.* 30: 26-28.

Melchers, G. 1960. Haploide Blütenpflanzen als Material der Mutationszüchtung. *Züchter* 30: 129-134.

Mendel, G. 1865 (1866). Versuche über Pflanzen-Hybriden. *Verhandlungen des Naturforschenden Vereines, Brünn* 4: 3-47.

Metzger, J. 1841. *Die Getreidearten und Wiesengräser in botanischer und ökonomischer Hinsicht. Akad. Verl.buchh.* F. Winter, Heidelberg, Deutschland.

Mikkelson, T. R., Anderson, B., Jørgensen, R. B. 1996. The Risk of Transgene Spread. *Nature* 380: 31.

Miller, P. 1768. *The Gardeners Dictionary.* C. Rivington, London.

Miller, T. E., Hutchinson, J. and Chapman V. 1982. Investigation of a Preferentially Transmitted *Aegilops sharonensis* Chromosome in Wheat. *Theor. Appl. Genet.* 61: 27-33.

Moldenke, H. N. and Moldenke, A. C. 1952. *Plants of the Bible.* Chronica Botanica Co., Waltham.

Monaghan, F. and Corcos, A. 1986. Tschermak: A Non-Discover of Mendelism. I. An Historical Note. *J. Heredity* 77: 468-469.

Monaghan, F. and Corcos, A. 1987. Tschermak: A Non-Discover of Mendelism. II. A Critique. *J. Heredity* 78: 208-210.

Moon, M. H. 1958. The Botanical Explorations of Liberty Hyde Bailey. 1. China. *Baileya* 6: 1-9.

Morel, G. and Martin, C. 1952. Guérison de dahlias atteints d'une maladie à virus. *C. R. Acad. Sci. Paris* 235: 1324-1325.

Morel, G. and Wetmore, R. H. 1951. Tissue Culture of Monocotyledons. *Amer. J. Bot.* 38: 138-140.

Morgan, T. H., Sturtevant, A. H., Muller, H. J. and Bridges, C. B. 1915. *The Mechanism of Mendelian Heredity*. H. Holt Press, New York.

Mosella, L. C., Signoret, P. A. and Jonard, R. 1980. Sur la mise au point de techniques de microgreffage d'apex en vue de l'élimination de deux types de particules virales chez le pêcher (*Prunus persica* Batsch). *C. R. Acad. Sci. Paris* 290: 287-290.

Muller, H. J. 1927. Artificial Transmutation of the Gene. *Science* 66: 84-87.

Munk, L., 1972. Improvement of Nutritional Value in Cereals. *Hereditas* 72: 11-128.

Müntzing, A. 1937. Polyploidy from Twin Seedlings. *Cytologia Fujii Jubilaei Vol.* 211-227.

Murai, N. and Kemp, J. D. 1982. T-DNA of pTi-15955 from *Agrobacterium tumefaciens* Is Transcribed into a Minimum of Seven Polyadenylated RNAs in a Sunflower Crown Gall Tumor. *Nucleic Acids Res.* 11: 1679-1689.

Murashige, T. and Skoog, F. 1962. A Revised Medium for Rapid Growth and Bioassays with Tobacco Tissue Cultures. *Physiol. Plants* 15: 473-497.

Nagata, T. and Bajaj, Y. P. S. 2001. *Somatic Hybridization in Crop Improvement*. Springer Verl. Berlin, Heidelberg, New York.

Nägeli, C. and Peter, A. 1885. *Die Hieracien Mittel-Europas*. R. Oldenbourg, München, Germany.

Nakajima, K. 1991. *Biotechnology for Crop Improvement and Production in Japan*. Regional Expert Consultation on the Role of Biotechnology in Crop Production, FAO Regional Office for Asia and the Pacific, Bangkok.

Naudin, C. 1863. Noevelles recherches sur l'hybridité dans les végétaux. *Ann. Soc. Natur. Botanique* 19: 180-203.

Navarro, L. and Juarez, J. 1977. Tissue Culture Techniques Used in Spain to Recover Virus-Free Citrus Plants. *Acta Hortic.* 78: 425-453.

Nebel, B. R. 1937. Mechanism of Polyploids through Colchicine. *Nature* 140: 1101.

Neyman, J. and Pearson, E. S. 1933. On the Problem of the Most Efficient tests of Statistical Hypotheses. *Philos. Trans. Roy. Soc.* 231: 289-337.

Nilsson-Ehle, H. 1908. Einige Ergebnisse von Kreuzungen bei Hafer und Weizen. *Botan. Notizen*: 257-294.

Nitsch, J. P. 1951. Growth and Development In Vitro of Excises Ovaries. *Amer. J. Bot.* 38: 566-571.

Nobbe, F. 1876. *Handbuch der Samenkunde*. Verl. Wiegandt, Berlin (Germany).

Oehlkers, F. 1943. Die Auslösung von Chromosomen-Mutationen in der Meiosis durch Einwirkung von Chemikalien. *Z. ind. Abst. Verer.-lehre* 81: 313-341.

Olby, R. C. 1966. *Origins of Mendelism*. Schocken Books, New York.

Olsson, G. 1986. *Svalöf 1886-1986—Research and Results in Plant Breeding.* LTS Vörlag, Stockholm: 290.

O'Mara, J. G. 1947. The Substitution of a Specific *Secale cereale* Chromosome for a Specific *Triticum aestivum* Chromosome. *Genetics* 32: 99-100.

Orel, V. 1996. *Gregor Mendel: The First Geneticist.* Oxford Univ. Press, Oxford.

Ortiz, R. 1998. Critical Role of Plant Biotechnology for the Genetic Improvement of Food Crops: Perspectives for the Next Millennium. *Electronic J. Biotechnol.* ISSN: 0717-3458.

Orton, W. A. 1900. The Wilt Disease of Cotton. *USDA Div. Veg. Physiol. Path. Bull.* 27.

Parthasarathy, N. and Rajan, S. S. 1953. Studies on Fertility of Autotetraploids of *Brassica campestris* var. *toria. Euphytica* 2: 25-36.

Paterson, A. H. 1996. *Genome Mapping in Plants.* Academic Press, Inc. & R. G. Landes Co., New York, Austin.

Paterson, A. H. 1997. *Molecular Dissection of Complex Traits.* CRC Press, Inc., Boca Raton.

Patterson, H. D. 1952. The Construction of Balanced Designs for Experiments Involving Sequences of Treatments. *Biometrika* 39: 32-48.

Pearson, E. S. 1931. Analysis of Variance in Cases of Non-Normal Variation. *Biometrika* 23: 114-133.

Pearson, E. S. 1937. Some Aspects of the Problem of Randomization. *Biometrika* 29: 53-64.

Pearson, K. 1900. Mathematical Contribution Theory of Evolution: On the Law of Reversion. *Proc. Roy. Soc.,* London.

Percival, J. 1921. *The Wheat Plant. A Monograph.* Duchworth Co., London.

Petry, E. 1922. Zur Kenntnis der Bedingungen der Biologischen Wirkungen der Röntgenstrahlen. *Biochem. Zeitschr.* 128: 326-353.

Plackett, R. L. and Burman, J. P. 1946. The Design of Optimum Multifactorial Experiments. *Biometrika* 33: 305-325.

Planck, J. E. van der 1963. *Plant Disease Epidemics and Control.* Academic Press, New York.

Powell-Abel, P., Nelson, R., De, B., Hoffmann, N., Rogers, G., Fraley, T. and Beachy, R. N. 1986. Delay of Disease Development in Transgenic Plants that Express the Tobacco Mosaic Virus Coat Protein Gene. *Science* 232: 738-743.

Punnett, R. C. 1907. *Mendelism.* 2nd ed., Bowes and Bowes, Cambridge, UK.

Quételet, L. A. 1871. *Anthropométrie ou Mesure des différentes facultés de l'homme.* C. Muquardt, Paris, France.

Rajaram, S., Maan, C. E., Ortiz-Ferrara, A. and Mujeeb-Kazi, A. 1983. Adaptation, Stability and High Yield Potential of Certain 1B/1R CIMMYT Wheats. *Proc. 6th Int. Wheat Genet. Symp.,* Kyoto, 613-621.

Randolph, L. F. 1932. Some Effects of High Temperature on Polyploidy and Other Variations in Maize. *Proc. Natl. Acad. Sci. USA* 18: 222-229.

Rao, S. R. and Ravishankar, G. A. 2000. Vanilla Flavour: Production by Conventional and Biotechnological Routes. *J. Food Sci.* 80: 289-304.

Redenbaugh, K. 1993. *Synseeds. Applications of Synthetic Seeds to Crop Improvement.* CRC Press, Boca Raton.

Reed, C. A. 1977. *The Origins of Agriculture.* Mouton Publ., The Hague, Netherlands.

Reed, H. 1942. *A Short History of the Plant Sciences.* Ronald Press, Waltham.

Reinert, J. 1959. Über die Kontrolle der Morphogenese und die Induction von adventiven Embryonem an Gewebekulturen von Karotten. *Planta* 53: 318-333.

Rhoades, M. M. 1931. Cytoplasmic Inheritance of Male Sterility in *Zea mays. Science* 73: 340-341.

Rhoades, M. M. 1955. *The Cytogenetic of Maize.* Academic Press, New York.

Richey, F. D. 1920. Corn Breeding. *Adv. Genet.* 3: 159-192.

Richey, F. D. 1946. Hybrid Vigor and Corn Breeding. *J. Amer. Soc. Agron.* 38: 833-841.

Rick, C. M. and Butler, L. 1956. Cytogenetics of the Tomato. *Adv. Genet.* 8: 267-382.

Rick, C. M. and Fobes, J. F. 1974. Association of an Allozyme with Nematode Resistance. *Rep. Tomato Genet. Coop.* 24: 25.

Ridley, M. 1984. The Horticultural Abbot of Brunn. *New Sci.* 5: 24-27.

Rifkin, J. 1998. *The Biotech Century.* Victor Gollancz, London.

Riley, R. 1960. The Diploidisation of Polyploid Wheat. *Heredity* 15: 407-429.

Riley, R. and Chapman, V. 1958. The Production and Phenotypes of Wheat-rye Chromosome Addition Lines. *Heredity* 12: 301-315.

Riley, R., Chapman, V. and Johnson, R. 1968. Introduction of Yellow Rust Resistance of *Aegilops comosa* into Wheat by Genetically Induced Homoeologous Recombination. *Nature* 217: 383-384.

Rimpau, W. 1877. *Züchtung neuer Getreidearten.* Landw. Jahrb., Deutschland.

Rimpau, W. 1891. *Kreuzungsprodukte landwirtschaftlicher Kulturpflanzen.* Verl. P. Parey, Berlin.

Rimpau, W. 1899. Monstrositäten am Roggen. *Dtsch. Landw. Presse* 26: 878-901.

Röbbelen, G. 2000 and 2002. Biographisches Lexikon zur Geschichte der Pflanzenzüchtung, Band 1 und 2, Gesell. f. Pflanzenz., Göttingen, Germany.

Roberts, H. F. 1929. *Plant Hybridization Before Mendel.* Univ. Press, Princeton.

Rodgers, A. D., III. 1965. *A Story of American Plant Sciences.* Hafner Press, New York.

Rohde, E. S. 1927. *Garden—Craft in the Bible.* Herbert Jenkins, London.

Ruckenbauer, P. 2000. E. von Tschermak-Seysenegg and the Austrian Contribution to Plant Breeding. *Vortr. Pflanzenzüchtung* 48: 31-46.

Rümker, K. von. 1889. *Anleitung zur Getreidezüchtung auf wissenschaftlicher und praktischer Grundlage.* P. Parey Verl., Berlin: 97-101.

Sageret, M. 1826. Considération sur la Production des Hybrides, des Variantes et des Variétés en général, et sur celles de la famille des Cucurbitacées rn particular. *Ann. Soc. Natur. Paris* 8: 294-314.

Sakamura, T. 1918. Kurze Mitteilung über die Chromosomenzahlen und die Verwandtschaftsverhältnisse der *Triticum*-Arten. *Botan. Mag., Tokyo* 32: 151-154.

Sanborn, J. W. 1890. Indian Corn. *Rept. Main Dept. Agric.* 33: 54-121.

Sapehin, A. A. 1935. X-ray Mutants in Soft Wheat. *Botan. Zhur., USSR* 20: 3-9.

Sauer, C. 1969. *Agricultural Origins and Dispersals.* MIT Press, Cambridge and London.

Sax, K. 1922. Sterility in Wheat Hybrids. II. Chromosome Behavior in Partially Sterile Hybrids. *Genetics* 7: 513-552.

Sax, K. 1923. The Association of Size Difference with Seed-Coat Pattern and Pigmentation in *Phaseolus vulgaris*. *Genetics* 8: 552-560.

Schaller, C. W. and Wiebe, G. A. 1952. Sources of Resistance to Net Blotch of Barley. *Agron. J.* 44: 334-336.

Schell, J., Montagu, M. van, Holsters, M., Zambryski, P., Joos, H., Inze, V., Herrera-Estrella, L., Depicker, A., Block, M. de, Caplan, A., Dhaese, P., Haute, E. van, Hernalsteens, J.-P., Greve, H. de, Leemans, J., Deblaere, R., Willmitzer, L., Schroder, J. and Otten, L. 1983. Ti Plasmids as Experimental Gene Vectors for Plants. *Adv. Gene Techn.*: Molecular Genetics of Plants and Animals. *Miami Winter Symp.* 20: 191-209.

Scheller, J., Guhrs, K. H., Grosse, F. and Conrad, U. 2001. Production of Spider Silk Proteins in Tobacco and Potato. *Nature Biotechnol.* 19: 573-577.

Schena, M., Shalon, D., Davis, R. W. and Brown, O. P. 1995. Quantitative Monitoring of Gene Expression Patterns with a Complementary DNA Micro Array. *Science* 270: 467-470.

Schlegel, R. 1976. The Relationship Between Meiosis and Fertility in Autotetraploid Rye, *Secale cereale* L. *Tag. Ber. Akad. d. Landwirtschaftswissenschaften, Berlin, Germany* 143: 31-36.

Schlegel, R. 1996. Triticale—Today and Tomorrow. In H. Guedes-Pinto, N. Darvey & V. Carnide (eds.), *Triticale Today and Tomorrow, Developments in Plant Breeding, Vol. 5* pp. 21-32, Kluwer Acad. Publ.

Schlegel, R. 2005. *Rye (Secale cereale L.)—A Younger Crop Plant with Bright Future*. CRC Press, Boca Raton.

Schlegel, R. and Korzun, V. 1997. About the Origin of 1RS.1BL Wheat-rye Chromosome Translocations from Germany. *Plant Breed.* 116: 537-540.

Schlegel, R., Kynast, R., Schwarzacher, T., Roemheld, V. and Walter, A. 1993. Mapping of Genes for Copper Efficiency in Rye and the Relationship between Copper and Iron Efficiency. *Plant and Soil* 154: 61-65.

Schlegel, R. and Meinel, A. 1994. A Quantitative Trait Locus (QTL) on Chromosome Arm 1RS of Rye and Its Effect on Yield Performance of Hexaploid Wheat. *Cer. Res. Comm.* 22: 7-13.

Schlegel, R., Melz, G. and Korzun, V. 1997. Genes, Marker and Linkage Data of Rye (*Secale cereale* L.). 5th updated inventory. *Euphytica* 101: 23-67.

Schlegel, R. and Mettin, D. 1975. Studies of Valence Crosses in Rye (*Secale cereale* L.). IV. The Relationship Between Meiosis and Fertility in Tetraploid Hybrids. *Biol. Zbl.* 94: 295-302.

Schlegel, R., Vahl, U. and Müller, G. 1994. A Compiled List of Wheats Carrying Homoeologous Group 1 Wheat-Rye Translocations and Substitutions. *Ann. Wheat Newslett., USA* 40: 105-117.

Schleiden, M. J. 1938. Beiträge zur Phytogenesis. *Arch. Anat. Physiol. Wiss. Med.* 5.

Schmalz, H. 1969. *Planzenzüchtung, Entwicklung–Stand–Aufgaben.* Deut. Landwirtschaftsverl. Berlin, Deutschland.

Schwanitz, F. 1967. *Die Evolution der Kulturpflanzen.* Bayr. Landw. Verl., München, Basel, Wien.

Schwann, T. 1839. *Mikroskopische Untersuchungen über die Übereinstimmung in der Struktur und im Wachstum der Thiere und Pflanzen.* Verl. W. Engelmann, Leipzig, Deutschland.

Sears, E. R. 1959. Aneuploids in Common Wheat. *Proc. 1st Int. Wheat Genet. Symp. Manitoba,* 221-228.

Sears, E. R. 1961. Identification of the Wheat Chromosome Carrying Leaf Rust Resistance from *Aegilops umbellulata. Wheat Inf. Serv.* 12: 12.

Sears, E. R. 1965. Nulli-Tetrasomic Combinations in Hexaploid Wheat. *Suppl. J. Heredity* 20: 29-45.

Sears, E. R. and Okamoto, M. 1958. Intergenomic Chromosome Relationships in Hexaploid Wheat. *Proc. 10th Int. Congr. Genet.* 2: 258-259.

Sebesta, E. E. and Wood, E. A. 1978. Transfer of Greenbug Resistance from Rye to Wheat with X-rays. *Agron. Abstr.* 61-62.

Seibert, M. 1976. Shoot Initiation from Carnation Shoot Apices Frozen to $-196°C$. *Science* 191: 1178-1179.

Semal, J. and Lepoivre, P. 1992. Biotechnologie et Agriculture: Impact et Perspectives. *Cahiers Agriculteurs* 13: 153-162.

Sengbusch, R. von. 1930. Bitterstoffarme Lupinen. I. *Züchter* 2: 1-7.

Seton, A. 1824. On the Variation in the Color of Peas from Cross-Impregnation. *Transact. Hort. Soc, London* 5: 236.

Shamel, A. D. 1905. *The Effect of Inbreeding in Plants.* Yearbook USDA, Washington, DC: 377-392.

Shepherd, K. W. and Islam, A. K. M. R. 1988. Fourth Compendium of Wheat-alien Chromosome Lines. *Proc. 7th Int. Wheat Genet. Symp.,* Cambridge, 1373-1395.

Shirreff, P. 1873. *Improvement of the Cereals and an Essay on the Wheat-fly.* Print for Private Circulation b. W. Blackwood & Sons, Edinburgh, London.

Shull, G. H. 1908. The Composition of a Field of Maize. *Rept. Amer. Breeders' Assoc.* 4: 296-301.

Shull, G. H. 1910. Hybridization Methods in Corn Breeding. *Amer. Breeders' Mag.* 1: 98-107.

Shull, G. H. 1912. The Influence of Inbreeding on Vigor of *Hydatina senta. Biol. Bull.* 24: 1-13.

Shull, G. H. 1922. Über die Heterozygotie mit Rücksicht auf den praktischen Zuchterfolg. *Beitr. Pflanzenzüchtung* 5: 134-158.

Shull, G. H. 1948. What is "Heterosis"? *Genetics* 33: 439-446.

Shull, G. H. 1952. *Beginning of the Heterosis Concept.* Iowa State College Press, Ames: 14-48.

Sijmons, P. C., Dekker, B. M., Schrammeijer. B., Verwoerd, T. C., van den Elzen, P. J., Hoekema, A. 1990. Production of Correctly Processed Human Serum Albumin in Transgenic Plants. *Biotechnology* 8: 217-221.

Sinclair, T. R., Purcell, L. C. and Sneller, C. H. 2004. Crop Transformation and the Challenge to Increase Yield Potential. *Trends Plant Sci.* 9: 70-75.

Singer, E., Holmyard, E. J. and Hall, A. R. 1954. *A History of Technology. Vol. 1. From Early Times to the Fall of Ancient Empires.* Oxford Univ. Press, London.

Sjodin, C. and Glimelius, K. 1989. Transfer of Resistance against *Phoma lingam* to *Brassica napus* by Asymmetric Somatic Hybridization Combined with Toxin Selection. *Theor. Appl. Genet.* 78: 513-520.

Smith, B. D. 1995. *The Emergence of Agriculture. Scientific American Library*, W. H. Freeman & Co., New York.

Smith, C. J. S., Watson, C. F., Ray, J., Bird, C. R., Morris, P. C., Schuch, W. and Grierson, D. 1988. Antisense RNA Inhibition of Polygalacturonase Gene Expression in Transgenic Tomatoes. *Nature* 334: 724-726.

Smith, H. H. 1943. Effects of Genomic Balance, Polyploidy, and Single Extra Chromosomes on Size in *Nicotiana. Genetics* 28: 227-236.

Smith, M. E. 1996, *The Aztecs.* Blackwell, Oxford, UK.

Snape, J. W., Simpson, E., Parker B. B., Friedt, W. and Foroughi-Wehr, B. 1986. Criteria for the Selection and Use of Double Haploid Systems in Cereal Breeding Programmes. In W. Horn, C. J. Jensen, W. Odenbach and Schieder, O., W. de Gruyter Verl. (eds.), *Genetic Manipulation in Plant Breeding. Proc. Int. Symp. Eucarpia* pp. 217-229, Berlin.

Solbrig, O. T. and Solbrig, D. J. 1994. *Farming and Crops in Human Affairs.* Island Press. Washington, DC.

Song, R. and Messing, J. 2003. Gene Expression of a Gene Family in Maize Based on Non-collinear Haplotypes. *Proc. Natl. Acad. Sci. USA* 100: 9055-9060.

Sosna, M. 1966. G. Mendel Memorial Symposium 1865-1965. *Proc. Symp., Brno,* Aug. 4-7, 1965. *Acad. Publ. House Czechoslovak Acad. Sci.* Prague.

Spillman, W. J. 1901. Quantitative Studies on the Transmission of Parental Characters of Hybrid Offspring. *Proc. 15th Ann. Convent Ass. Amer. Agric. Coll. Exper. Stat., Washington. Off. Exper. Stat. Bull.* 115.

Sprague, G. F. and Jenkins, M. T. 1943. A Comparison of Synthetic Varieties, Multiple Crosses and Double Crosses in Corn. *J. Amer. Soc. Agron.* 35: 137-147.

Sprague, G. F. and Tatum, L. A. 1942. General vs. Specific Combining Ability in Single Crosses of Corn. *J. Amer. Soc. Agron.* 34: 923-932.

Sprengel, C. K. 1793. *Das entdeckte Geheimnis der Natur im Bau und in der Befruchtung der Blumen.* Ostwalds Klassiker der exakten Wissenschaften, Leipzig.

Stadler, L. J. 1928. Genetic Effects of X-rays in Barley. *Proc. Natl. Acad. Sci. USA* 68: 186-187.

Stadler, L. J. 1929. Chromosome Number and the Mutation Rate in *Avena* and *Triticum. Proc. Natl. Acad. Sci. USA* 15: 876-881.

Stein, E. 1922. Über den Einfluss von Radiumstrahlung auf Anthirrhinum. *Z. ind. Abst. Verb.-lehre* 29: 1-15.

Stern, C. 1931. Zytologisch-genetische Untersuchungen als Beweise für die Morgansche Theorie des Faktorenaustauschs. *Biol. Zbl.* 51: 547-587.

Stern, C. and Sherwood, E. R. 1966. *The Origin of Genetics. A Mendel Source Book.* W. H. Freeman, San Francisco.

Steward, F. C. Mapes, M. O. and Mears, K. 1958. Growth and Organized Development of Cultured Cells. II. Organization in Cultures Grown from Freely Suspended Cells. *Amer. J. Bot.* 45: 705-708.

Strasburger, E. 1910. Chromosomenzahl. *Flora* 100: 398-444.

Stubbe, H., 1965. *Kurze Geschichte der Genetik bis zur Wiederentdeckung der Vererbungsregeln Gregor Mendels.* G. Fischer Verl., Jena.

Stuber, C. W., Polacco, M. and Senior, M. L. 1999. Synergy of Empirical Breeding, Marker-Assisted Selection, and Genomics to Increase Crop Yield Potential. *Crop Sci.* 39: 1571-1583.

"Student." 1938. Comparison Between Balanced and Random Arrangements of Field Plots. *Biometrika* 29: 363-379.

Sturtevant, A. H. 1913. The Linear Arrangement of Six Sex-Linked Factors in *Drosophila,* as Shown by Their Mode of Association. *J. Exper. Zool.* 14: 43-59.

Sturtevant, A. H. 1917. Crossing Over without Chiasmatype? *Genetics* 2: 301-304.

Sturtevant, A. H. 1965. *A History of Genetics.* Harper & Row, New York.

Sukekiyo, Y., Ogura, H., Kimura, Y. and Itoh, R. 1989. Development of a New Rice Variety Hatsuyume by Protoplast Breeding. *Proc. 6th Inter. Congr.* SABRAO: 497-500.

Suneson, C. A. 1956. An Evolutionary Plant Breeding Method. *Agron. J.* 48: 188-190.

Suneson, C. A. and Wiebe, G. A. 1942. Survival of Barley and Wheat Varieties in Mixtures. *J. Amer. Soc. Agron.* 34: 1052-1056.

Sutton, W. S. 1902. On the Morphology of the Chromosome Group in *Brachystola magna. Biol. Bull. Marin. Biol. Labor.* 4: 1-16.

Swingle, W. T. 1913. *The Date Palm and Its Utilization in the Southwestern States.* Bureau Plant Industry, USDA, USA.

Swingle, W. T. and Webber, H. J. 1897. Hybrids and Their Utilization in Plant Breeding. *Yearbook U.S. Dep. Agric.* 383-420.

Sybenga, J., 1964. The Use of Chromosomal Aberrations in the Autopolyploidization of Autopolyploids. *Proc. Symp. IAEA/FAO,* Rome, 741-749.

Tacker, C. O., Mason, H. S., Losonsky, G., Clements, J. D., Levine, M. M. and Arntzen, C. J. 1998. Immunogenecity in Humans of a Recombinant Bacterial Antigen Delivered in a Transgenic Potato. *Nature Medicine* 4: 607-609.

Tammes, T. 1924. Das genotypische Verhältnis zwischen dem wilden *Linum angustifolium* and dem Kulturlein, *L. usitatissimum. Genetica* 5: 61-76.

Tanksley, S. D., Medino-Filho, H. and Rick, C. M. 1982. Use of Naturally Occurring Enzyme Variation to Detect and Map Genes Controlling Quantitative Traits in an Interspecific Backcross of Tomato. *Heredity* 49: 11-25.

Tanksley, S. D. and Orton, T. J. 1983. *Isozymes in Plant Genetics and Breeding.* Elsevier, New York.

Tanksley, S. D., Rick, C. M. and Vallejos, C. E. 1984. Tight Linkage between a Nuclear Male-sterile Locus and an Enzyme Marker in Tomato. *Theor. Appl. Genet.* 68: 109-113.

Thacker, C. 1979. *The History of Gardens.* Univ. of California Press, Berkeley.

Tschermak, E. von 1898. Gemüsesamenzucht in Deutschland. *Wiener Landwirt. Zeitg.* 42: 343-344.

Tschermak, E. von 1900. Über künstliche Kreuzung bei *Pisum sativum. Zeitschr. landwirt. Versuchswesen in Österreich* 3: 465-555.

Tschermak, E. von 1901. Über Züchtung neuer Getreiderassen mittels künstlicher Kreuzung. I. Kritisch-historische Betrachtungen. *Zeitschr. landwirt. Versuchswesen in Österreich* 4: 1029-1060.

Tschermak, E. von 1906. Über Züchtung neuer Getreiderassen mittels künstlicher Kreuzung. II. Kreuzungsstudien am Roggen. *Zeitschr. landwirt. Versuchswesen in Österreich* 9: 699-743.

Tschermak, E. von 1932. Über einige Blütenanomalien bei Primeln und ihre Vererbungsweise. *Biologica Generalis* 8: 337-350.

Tschermak, E. von 1951. The Rediscovery of Gregor Mendel's Work. *J. Heredity* 42: 163-171.

Tschermak, E. von und Bleier, H. 1926. Über fruchtbare *Aegilops*-Weizenbastarde. *Ber. Deutsch. Bot. Gesell.* 44: 110-132.

Tschermak, E. von und Rümker, K. von 1910. *Reisebericht über landwirtschaftliche Studien in Nord-Amerika mit besonderer Berücksichtigung der Pflanzenzüchtung.* Paul Parey Verlag, Berlin.

Tsuchiya, T. and Gupta, P. K. 1991. *Chromosome Engineering in Plants. A + B.* Elsevier Amsterdam, Oxford, New York, Tokyo.

Tysdal, H. M., Kiesselbach, T. A. and Westover, H. L. 1942. Alfalfa Breeding. *Univ. Nebraska Agric. Exp. Sta. Res. Bull.* 124.

Uauy, C., Distelfeld, A., Fahima, T., Blechl, A. and Dubcovsky, J. 2006. A NAC Gene Regulating Senescence Improves Grain Protein, Zinc, and Iron Content in Wheat. *Science* 314: 1298-1301.

Vasil, V. and Hildebrandt, A. C. 1965. Growth and Tissue Formation from Single, Isolated Tobacco Cells in Microculture. *Science* 147: 1454-1455.

Vavilov, N. I. 1926. Studies on the Origin of Cultivated Plants. *Inst. Bot. Appl. Amelior. Plants,* Leningrad.

Vavilov, N. I. 1928. Geographische Zentren unserer Kulturpflanzen (Geographic centers of our crop plants). *Zeitschr. ind. Abst. Vererb.-lehre, Suppl.* 1: 342-369.

Venn, J. 1888. Cambridge Anthropometry. *J. Anthropolog. Institute* 18: 140-154.

Vries, H. de, 1900. Das Spaltungsgesetz der Bastarde. *Ber. der Deutsch. Bot. Gesell.* 18: 83-90.

Vries, H. de 1901. *Die Mutationstheorie.* Verl. Veit & Co.

Vries, H. de. 1903. *Die Mutationstheorie* Vol. 2, Veit & Co., Leipzig, Germany.

Walker, W. 1979. *All the Plants of the Bible.* Doubleday & Company, Inc. Garden City, New York.

Watson, J. D. and Crick, F. H. C. 1953. Molecular Structure of Nucleic Acids. *Nature* 171: 737-738.

Webber, H. J. 1905. Explanation of Mendel's Law of Hybrids. *Proc. Amer. Breed. Assoc.* 1: 138-143.

Weinberg, W. R. 1909. Über Vererbungsgesetze beim Menschen. *Z. ind. Abst.-Vererb.-lehre* 1: 377-392.

Welch, B. L. 1937. On the z-Test in Randomized Blocks and Latin Squares. *Biometrika* 29: 21-52.

Wellensiek, S. J. 1947. Rational Methods for Breeding Cross-fertilizers. *Mededel. Landbouwhogeschool Wageningen* 48: 227-262.

White, P. R. 1934. Potentially Unlimited Growth of Excised Tomato Root Tips in a Liquid Medium. *Plant Physiol.* 9: 585-600.

Wiegemann, A. F. 1828. *Über die Bastarderzeugung im Pflanzenreiche.* Verl. F. Vieweg, Braunschweig, Deutschland.

Willcox, G. 1998. Archaeobotanical Evidence for the Beginnings of Agriculture in Southwest Asia. In A. Damania, et al. (eds.) *The Origins of Agriculture and Crop Domestication* pp. 25-38. ICARDA, Aleppo, Syria.

Wilson, W. 1900. *The Cell in Development and Heredity.* Macmillan Comp., New York.

Winge, O. 1917. The Chromosomes. Their Number and General Importance. *Comp. Rend. Trav. Lab. Carlsberg* 13: 131-275.

Winkler, H. 1916. Über die experimentelle Erzeugung von Pflanzen mit abweichenden Chromosomenzahlen. *Z. Bot.* 8: 417-424.

Wollny, E. 1885. *Saat und Pflege der landwirtschaftlichen Kulturpflanzen.* Verl. P. Parey, Berlin.

Wright, S. 1921. Systems of mating. I. The biometric relation between parent and offspring. *Genetics* 6: 111-123.

Wunderlich, G. 1951. Die Bedeutung Tschermaks für den österreichischen Getreidebau. *Zeitschr. Pflanzenzüchtung* 30: 478-483.

Xu, H., Swoboda, I., Bhalla, P. L., Sijbers, A., Chao, C., Ong, E., Hoeijmakers, J. H. J. and Singh, M. B. 1998. Human Nucleotide Excision Repair Gene ERCC1 Homologue in Plants and Its Preferential Expression in Male Germ Line Cells. *Plant J.* 13: 823-829.

Xu, H., Swoboda, I., Bhalla, P. L. and Singh, M. B. 1999. Male Gametic Cell Specific Gene Expression in Flowering Plants. *Proc. Natl. Acad. Sci. USA* 96: 2554-2558.

Yates, F. 1939. The Comparative Advantages of Systematic and Randomized Arrangements in the Design of Agricultural and Biological Experiments. *Biometrika* 30: 440-466.

Yates, F. 1951. The Influence of Statistical Methods for Research Workers on the Development of the Science of Statistics. *J. Amer. Statist. Assoc.* 46: 19-34.

Ye, X., Al-Babili, S., Kloeti, A., Zhang, J., Lucca, P., Beyer, P. and Potrykus, I. 2000. Engineering Provitamin A (ß-carotene) Biosynthetic Pathway into (Carotenoid-free) Rice Endosperm. *Science* 287: 303-305.

Zeller, F. J. 1973. 1B/1R Wheat-Rye Substitutions and Translocations. *Proc. 4th Int. Wheat Genet. Symp.,* Columbia, 209-221.

Zeller, F. J., Günzel, G., Fischbeck, G., Gertenkorn, P. and Weipert, D. 1982. Veränderungen der Backeigenschaften der Weizen-Roggen-Translokationssorten. *Getreide Mehl Brot* 36: 141-143.

Zeller, F. J. and Hsam, K. 1983. Broadening the Genetic Variability of Cultivated Wheat by Utilizing Rye Chromatin. *Proc. 6th Int. Wheat Genet. Symp.,* Kyoto, 161-173.

Zhang, X., Urry, D. W. and Daniell, H. 1996. Expression of an Environmentally Friendly Synthetic Protein-Based Polymer in Transgenic Tobacco Plants. *Plant Cell Rep.* 16: 174-179.

Zhao, B., Lin, X., Jesse Poland, J., Harold Trick, H., Leach, J. and Scot, H. 2005. A Maize Resistance Gene Functions against Bacterial Streak Disease in Rice. *Proc. Natl. Acad. Sci. USA* 102: 15383-15388.

Zimmermann, U. 1982. Electric Field-Mediated Fusion and Related Electrical Phenomena. *Biochim. Biophys. Acta* 694: 227-277.

Zirxel, C. 1935. *The Beginnings of Plant Hybridization.* Univ. of Pennsylvania Press, Philadelphia.

Index

Page numbers followed by the letter "e" indicate exhibits; "f" indicates figures; "p" indicates photos; and "t" indicates tables.

Concise Encyclopedia of Crop Improvement
© 2007 by The Haworth Press, Inc. All rights reserved.
doi:10.1300/5891_10